# Oxford International Primary Maths

# Workbook

# 2

Tony Cotton

OXFORD
UNIVERSITY PRESS

Great Clarendon Street, Oxford, OX2 6DP, United Kingdom

Oxford University Press is a department of the University of Oxford. It furthers the University's objective of excellence in research, scholarship, and education by publishing worldwide. Oxford is a registered trade mark of Oxford University Press in the UK and in certain other countries



British Library Cataloguing in Publication Data
Data available

978 0 19 836527 3

10 9 8 7

Paper used in the production of this book is a natural, recyclable product made from wood grown in sustainable forests. The manufacturing process conforms to the environmental regulations of the country of origin.

Printed in Great Britain by Ashford Print and Publishing Services, Gosport

**Acknowledgements**
The publishers would like to thank the following for permissions to use their photographs:

**Cover:** Saudi Desert Photos by TARIQ-M

Although we have made every effort to trace and contact all copyright holders before publication this has not been possible in all c ases. If notified, the publisher will rectify any errors or omissions at the earliest opportunity.

Links to third party websites are provided by Oxford in good faith and for information only. Oxford disclaims any responsibility for the materials contained in any third party website referenced in this work.

# Contents

# 1 Tens and Ones

## Introduction

Students need to understand the decimal nature of the number system. This unit introduces them to the idea that the number system is based on tens. Students will begin to understand that in a number such as 57:

- the 5 stands for 50 or 5 tens
- the 7 stands for 7 ones.

## Ways to help

At this stage it is still very helpful to count as often as possible. Starting at different numbers and counting forwards and backwards up to 100 allows students to hear the patterns and become secure with the names of the numbers. Looking for numbers in the environment and saying them aloud is also very helpful.

**Key Words**

count forwards; count backwards; how many tens?; how many ones?; one, two ...; ninety-nine; one hundred

# 1A Counting in tens and ones

Date: 27/1/20

## Discover

Write the missing numbers for each question.
Colour in the patterns on the 100 square.
Use a different colour for each question.
Say each number aloud as you write it and colour it.
The first one is done for you.

| 1 | 2 | 3 | 4 | 5 | 6 | 7 | 8 | 9 | 10 |
|---|---|---|---|---|---|---|---|---|---|
| 11 | 12 | 13 | 14 | 15 | 16 | 17 | 18 | 19 | 20 |
| 21 | 22 | 23 | 24 | 25 | 26 | 27 | 28 | 29 | 30 |
| 31 | 32 | 33 | 34 | 35 | 36 | 37 | 38 | 39 | 40 |
| 41 | 42 | 43 | 44 | 45 | 46 | 47 | 48 | 49 | 50 |
| 51 | 52 | 53 | 54 | 55 | 56 | 57 | 58 | 59 | 60 |
| 61 | 62 | 63 | 64 | 65 | 66 | 67 | 68 | 69 | 70 |
| 71 | 72 | 73 | 74 | 75 | 76 | 77 | 78 | 79 | 80 |
| 81 | 82 | 83 | 84 | 85 | 86 | 87 | 88 | 89 | 90 |
| 91 | 92 | 93 | 94 | 95 | 96 | 97 | 98 | 99 | 100 |

**1.** Count on in tens from 31.

31 | 41 | 51 | 61 | 71 | 81 | 91

**2.** Count on in tens from 18.

18 | 28 | 38 | 48 | 58 | 68 | 78 | 88 | 98

**3.** Count back in tens from 82.

82 | 72 | 62 | 52 | 42 | 32 | 22 | 12 | 2

**4.** Count back in tens from 65.

65 | 55 | 45 | 35 | 25 | 15 | 5

**5.** Count on in tens from 23.

23 | 33 | 43 | 53 | 63 | 73 | 83 | 93

**6.** Count on in tens from 49.

49 | 59 | 69 | 79 | 89 | 99

**7.** Count back in tens from 70.

70 | 60 | 50 | 40 | 30 | 20 | 10

**8.** Count back in tens from 97.

97 | 87 | 77 | 67 | 57 | 47 | 37 | 27 | 17 | 7

# 1A Counting in tens and ones

Date: 27/1/2020

## Explore

Practise counting on in ones and counting back in tens. Write the missing numbers. Say each number aloud as you write it.
The first one is done for you.

**1.**

| Count on in 1s | | | | | | | | | | |
|---|---|---|---|---|---|---|---|---|---|---|
| 71 | 72 | 73 | 74 | 75 | 76 | 77 | 78 | 79 | 80 | 81 |

| Count back in 10s | | | | |
|---|---|---|---|---|
| 71 | 61 | 51 | 41 | 31 |

**2.**

| Count on in 1s | | | | | | | | | |
|---|---|---|---|---|---|---|---|---|---|
| 83 | 84 | 85 | 86 | 87 | 88 | 89 | 90 | 91 | 92 |

| Count back in 10s | | | | |
|---|---|---|---|---|
| 83 | 73 | 63 | 53 | 43 | 33 |

**3.**

| Count on in 1s | | | | | | | | | |
|---|---|---|---|---|---|---|---|---|---|
| 65 | 66 | 67 | 68 | 69 | 70 | 71 | 72 | 73 | 74 | 75 |

| Count back in 10s | | | |
|---|---|---|---|
| 65 | 55 | 45 | 35 |

**4.**

| Count on in 1s | | | | | | | |
|---|---|---|---|---|---|---|---|
| 89 | 90 | 91 | 92 | 93 | 94 | 95 | 96 | 97 |

| Count back in 10s | | | | | |
|---|---|---|---|---|---|
| 89 | 79 | 69 | 59 | 49 | 39 | 29 |

**5.**

| Count on in 1s | | | | | | | | | | |
|---|---|---|---|---|---|---|---|---|---|---|
| 48 | 49 | 50 | 51 | 52 | 53 | 54 | 55 | 56 | 57 | 58 | 59 |

| Count back in 10s | | | |
|---|---|---|---|
| 48 | 38 | 28 | 18 |

**6.**

| Count on in 1s | | | | | | | | | | |
|---|---|---|---|---|---|---|---|---|---|---|
| 52 | 53 | 54 | 55 | 56 | 57 | 58 | 59 | 60 | 61 | 62 | 63 |

| Count back in 10s | | | |
|---|---|---|---|
| 52 | 42 | 32 | 22 |

**7.**

| Count on in 1s | | | | | | | | |
|---|---|---|---|---|---|---|---|---|
| 66 | 67 | 68 | 69 | 70 | 71 | 72 | 73 | 74 |

| Count back in 10s | | | | | | |
|---|---|---|---|---|---|---|
| 66 | 56 | 46 | 36 | 26 | 16 | 6 |

**8.**

| Count on in 1s | | | | | | | | | | | |
|---|---|---|---|---|---|---|---|---|---|---|---|
| 27 | 28 | 29 | 30 | 31 | 32 | 33 | 34 | 35 | 36 | 37 | 38 | 39 |

| Count back in 10s | | |
|---|---|---|
| 27 | 17 | 7 |

# 1B Digit values



On the place-value cards, write how many tens and how many ones make up each of the numbers below. The first one is done for you.

53 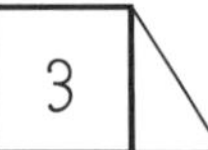 and 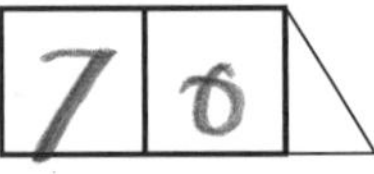

78  and 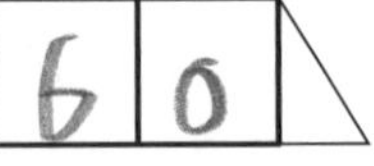

61 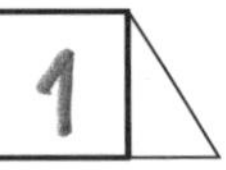 and 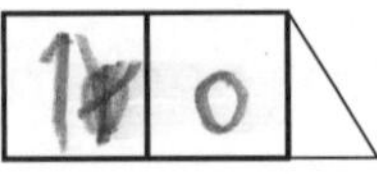

19 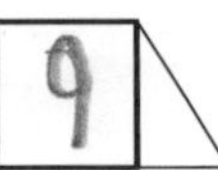 and 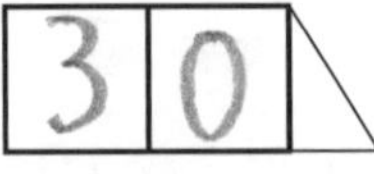

37 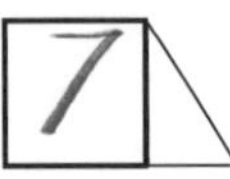 and 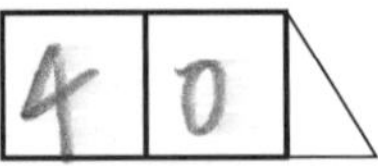

46 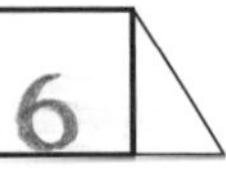 and 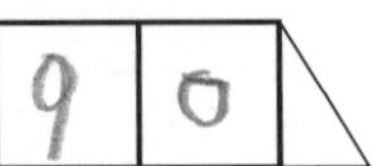

99 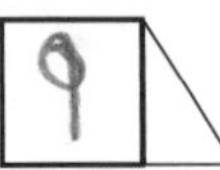 and 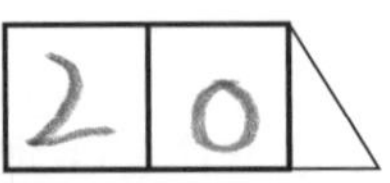

24 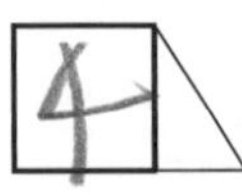 and 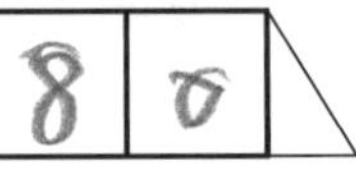

82  and 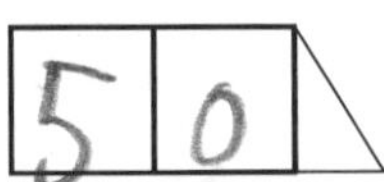

55  and 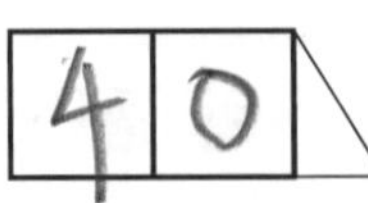

40 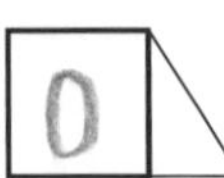 and 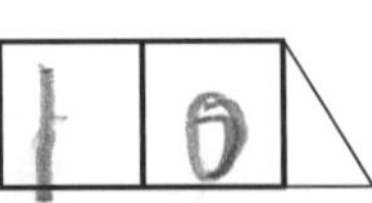

13 and

# 1B Digit values



Place-value counters are another way of showing how any 2-digit number is made up of tens and ones.

On the place-value cards write the number shown by the place-value counters.
Then write the number in the 'Tens' and 'Ones' boxes. Say each number aloud as you write it.

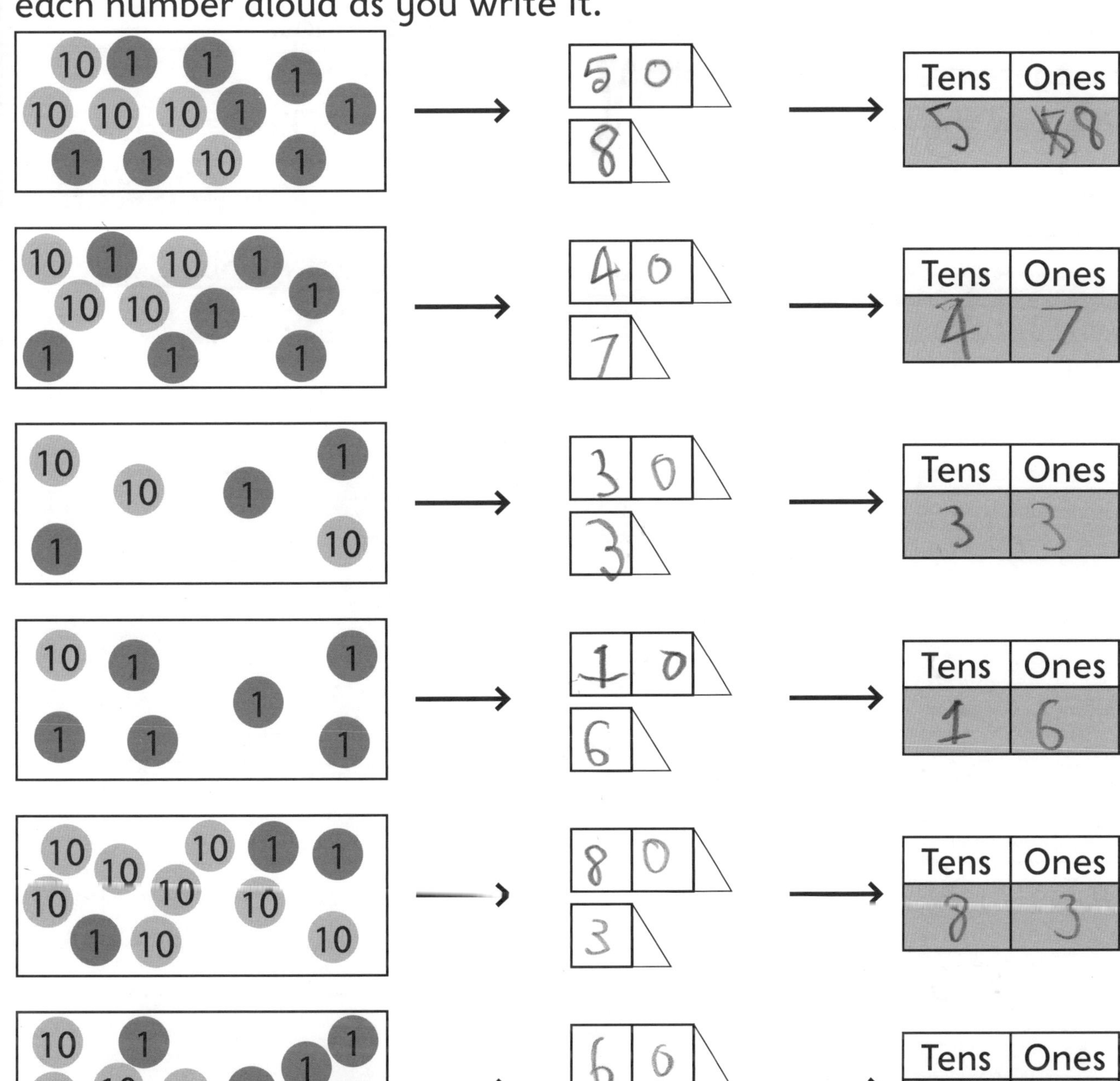

# 1C Estimating and counting



You will need to work with a friend to do this activity.

Count out 100 small objects such as dried beans or chickpeas. Make sure that you have exactly 100.

Take it in turns to pick up a handful of the beans and put them in a bowl. The other person should estimate how many there are to the nearest 10.

Then count the beans aloud together. Write exactly how many there are.

| Turn | Estimate (to the nearest 10) | Exact number |
| --- | --- | --- |
| Turn 1 |  |  |
| Turn 2 |  |  |
| Turn 3 |  |  |
| Turn 4 |  |  |
| Turn 5 |  |  |
| Turn 6 |  |  |

# 1C Estimating and counting

## Explore

For this activity you will need small objects such as dried
beans or chickpeas and five small containers.

Estimate the number of beans that you think you will
need to fill each container. Then fill each container with
beans and count them.

| Container | Estimate (to the nearest 10) | Exact number |
|---|---|---|
| Container 1 | | |
| Container 2 | | |
| Container 3 | | |
| Container 4 | | |
| Container 5 | | |

You will need a set of 0–9 digit cards for this activity.

Pick two cards, for example:

| 0 | 1 | 2 | 3 | 4 |
| 5 | 6 | 7 | 8 | 9 |

| 3 | 6 |

Make the largest number you can using these two digits.
This is your start number. Then practise counting back in ones and counting on in tens. Say each number aloud as you write it.
The first one is done for you.

**1.**

| Start number | Count back in 1s | | | | | | | Start number | Count on in 10s | |
| --- | --- | --- | --- | --- | --- | --- | --- | --- | --- | --- |
| 63 | 62 | 61 | 60 | 59 | 58 | 57 | 56 | 63 | 73 | 83 |

**2.**

| Start number | Count back in 1s | | | | | | | Start number | Count on in 10s | |
| --- | --- | --- | --- | --- | --- | --- | --- | --- | --- | --- |
| 24 | 23 | 22 | 21 | 20 | 19 | 17 | 16 | 24 | 34 | 44 |

**3.**

| Start number | Count back in 1s | | | | | | | Start number | Count on in 10s | |
| --- | --- | --- | --- | --- | --- | --- | --- | --- | --- | --- |
| 77 | 76 | 75 | 74 | 73 | 72 | 71 | 70 | 77 | 87 | 97 |

**4.**

| Start number | Count back in 1s | | | | | | | Start number | Count on in 10s | |
| --- | --- | --- | --- | --- | --- | --- | --- | --- | --- | --- |
| 88 | 87 | 86 | 85 | 84 | 83 | 82 | 81 | 88 | 98 | 108 |

**5.**

| Start number | Count back in 1s | | | | | | | Start number | Count on in 10s | |
| --- | --- | --- | --- | --- | --- | --- | --- | --- | --- | --- |
| 90 | 89 | 88 | 87 | 86 | 85 | 84 | 83 | 90 | 100 | 110 |

**6.**

| Start number | Count back in 1s | | | | | | | Start number | Count on in 10s | |
| --- | --- | --- | --- | --- | --- | --- | --- | --- | --- | --- |
| 33 | 32 | 31 | 30 | 29 | 28 | 27 | 26 | 33 | 43 | 45 |

**7.**

| Start number | Count back in 1s | | | | | | | Start number | Count on in 10s | |
| --- | --- | --- | --- | --- | --- | --- | --- | --- | --- | --- |
| 45 | 44 | 43 | 42 | 41 | 40 | 39 | 38 | 45 | 55 | 65 |

# 2 Number Patterns and Properties

## Introduction

This unit will help students to recognise number patterns, which they will use to help them in calculations. They will develop their understanding of odd and even numbers and use their knowledge of place value to compare and order numbers.

## Ways to help

Check that students understand that the tens number comes first when writing 2-digit numbers. Remind students that numbers are read from left to right in Hindu-Arabic (Western) script and that the number on the left has the highest value. For example, 26 is made from a 20 and a 6 and 14 is made from a 10 and a 4.

You can also help students to remember that the symbol > means 'greater than' and < means 'less than'.

**Key Words**

even; odd; double; half; multiple; between; greater than; less than; 1st and first; 2nd and second; 3rd and third and so on up to 10th and tenth

# 2A Odd and even



You will need two dice for this activity.

Roll both the dice and write the two 2-digit numbers you can make using the numbers on the dice.
Write both the numbers in the correct columns.
Repeat ten times.

Sometimes you will be able to make an even and an odd number. But sometimes both your numbers will be even and sometimes they will both be odd.

| Even | Odd |
| --- | --- |
| 14 | 41 |
|  |  |
|  |  |
|  |  |
|  |  |
|  |  |
|  |  |
|  |  |
|  |  |
|  |  |
|  |  |
|  |  |
|  |  |

## 2A Odd and even

### Explore

You will need two dice for this activity.

Roll both the dice and write the two 2-digit numbers
you can make using the numbers on the dice.
Write the numbers in the Carroll diagram.

|        | Less than 30 | Not less than 30 |
|--------|--------------|------------------|
| Even   |              | 56               |
| Odd    |              | 65               |

# 2B Doubles

## Discover

Can you remember your doubles? Complete the table.

Two are done for you.

| Number | Double |
| --- | --- |
| 1 | 2 |
| 2 | 4 |
| 3 | 6 |
| 4 | 8 |
| 5 | 10 |
| 6 | 12 |
| 7 | 14 |
| 8 | 16 |
| 9 | 18 |
| 10 | 20 |
| 15 | 30 |
| 20 | 40 |
| 25 | 50 |
| 50 | 100 |

**1.** Write the missing word in this sentence.

All doubles are __even__ numbers.

**2.** What do you notice about these doubles?

double 1 and double 10 ____________________

____________________

double 2 and double 20 ____________________

____________________

double 5 and double 50 ____________________

____________________

You can ask an adult to help you with the writing if you would like to.

# 2B Doubles

## Explore

You can work out bigger doubles by 'partitioning' into tens and ones. The first one is done for you.

**1.** 13 = [1 0] + [3]   Double each part: [2 0] + [6]

[20] + [6] = [26], so double 13 = [26]

**2.** 14 = [1 0] + [4]   Double each part: [2 0] + [8]

[20] + [8] = [28], so double 14 = [28]

**3.** 15 = [1 0] + [5]   Double each part: [2 0] + [10]

[20] + [10] = [30], so double 15 = [30]

**4.** 19 = [1 0] + [9]   Double each part: [2 0] + [18]

[20] + [18] = [38], so double 19 = [38]

**5.** 11 = [1 0] + [1]   Double each part: [2 0] + [2]

[20] + [2] = [22], so double 11 = [22]

**6.** 16 = [1 0] + [6]   Double each part: [2 0] + [12]

[20] + [12] = [32], so double 16 = [32]

# 2C Ordering and between

## Discover and Explore

Use the 100 square to help you find numbers **in between** the numbers you are given.

The first answer in each section is done for you.

| 1 | 2 | 3 | 4 | 5 | 6 | 7 | 8 | 9 | 10 |
|---|---|---|---|---|---|---|---|---|---|
| 11 | 12 | 13 | 14 | 15 | 16 | 17 | 18 | 19 | 20 |
| 21 | 22 | 23 | 24 | 25 | 26 | 27 | 28 | 29 | 30 |
| 31 | 32 | 33 | 34 | 35 | 36 | 37 | 38 | 39 | 40 |
| 41 | 42 | 43 | 44 | 45 | 46 | 47 | 48 | 49 | 50 |
| 51 | 52 | 53 | 54 | 55 | 56 | 57 | 58 | 59 | 60 |
| 61 | 62 | 63 | 64 | 65 | 66 | 67 | 68 | 69 | 70 |
| 71 | 72 | 73 | 74 | 75 | 76 | 77 | 78 | 79 | 80 |
| 81 | 82 | 83 | 84 | 85 | 86 | 87 | 88 | 89 | 90 |
| 91 | 92 | 93 | 94 | 95 | 96 | 97 | 98 | 99 | 100 |

## A number in between

| 14 | 16 | 17 |
|---|---|---|
| 23 |  28 | 29 |
| 31 |  32 | 35 |
| 46 |  49 | 50 |
| 58 |  62 | 63 |
| 62 |  63 | 66 |
| 70 |  74 | 75 |
| 78 | 80 | 81 |
| 82 | 83 | 84 |
| 93 | 98 | 99 |

## Two numbers in between

| 7 | 11 | 15 | 18 |
|---|---|---|---|
| 11 | 14 | 16 | 19 |
| 26 | 28 | 29 | 31 |
| 32 | 33 | 36 | 37 |
| 38 | 39 | 40 | 41 |
| 42 | 43 | 49 | 50 |
| 57 | 59 | 60 | 61 |
| 62 | 65 | 66 | 69 |
| 78 | 79 | 80 | 86 |
| 89 | 92 | 96 | 99 |

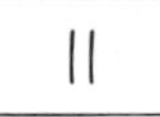

# 2D Less than, greater than

## Discover and Explore

You will need a set of 0–9 digit cards for this activity.

| 0 | 1 | 2 | 3 | 4 |
| 5 | 6 | 7 | 8 | 9 |

Turn over two cards and use them to make a 2-digit number.

Turn over two more cards and use them to make another 2-digit number.

Use > and < to write a number sentence using the two numbers you have made.

> **Remember**
> \> means 'is greater than'
> < means 'is less than'

| 26 | > | 13 |
| 59 | > | 56 |
| 500 | < | 550 |
| 79 | > | 52 |
| 46 | < | 200 |
| 15 | < | 42 |

### Discover

Follow the instructions to colour in the bead string. You will need red, blue, green, yellow and black crayons.

Colour the 5th bead in black.

Colour the 2nd bead in blue.

Colour the 13th bead in green.

Colour the 4th bead in red.

Colour the 16th bead in yellow.

Colour the 19th bead in red.

Colour the 1st bead in yellow.

Colour the 3rd bead in green.

Colour the 9th bead in red.

Now complete the repeating pattern by colouring in the rest of the beads.

Complete these sentences.

The 6th bead is ____________________.

The 9th bead is ____________________.

The 11th bead is ____________________.

The 20th bead is ____________________.

# 2E Ordinal numbers



Practise your ordinal numbers.

Colour in the 14th mouse.

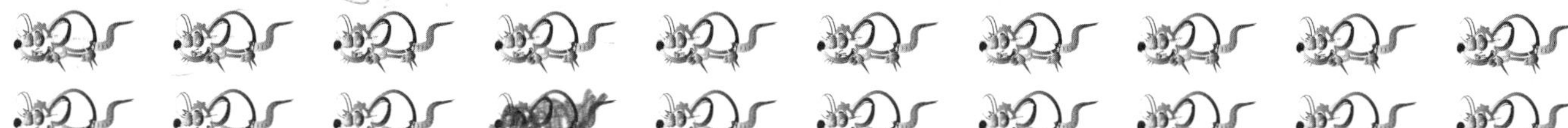

Colour in the 11th car.

Colour in the 17th butterfly.

Colour in the 17th person.

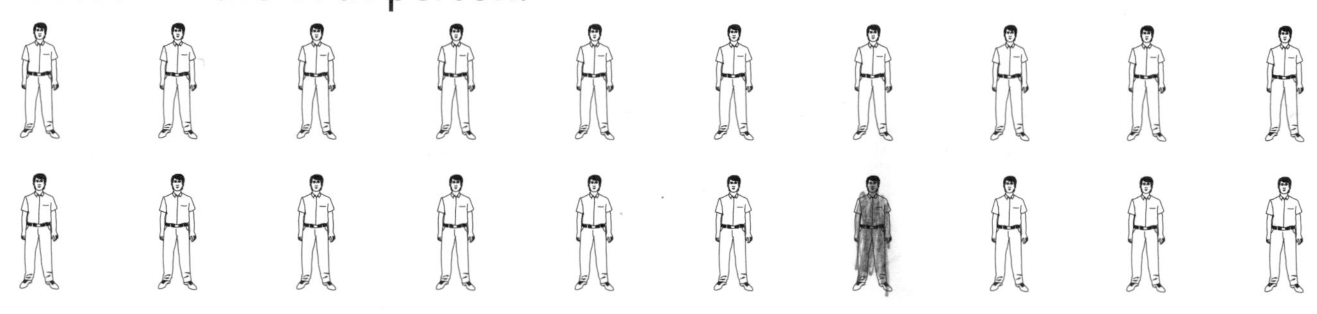

Colour in the 19th present.

Draw an arrow from each number to every number fact that is correct about that number. Use a different colour for each number if possible.

The first one is done for you.

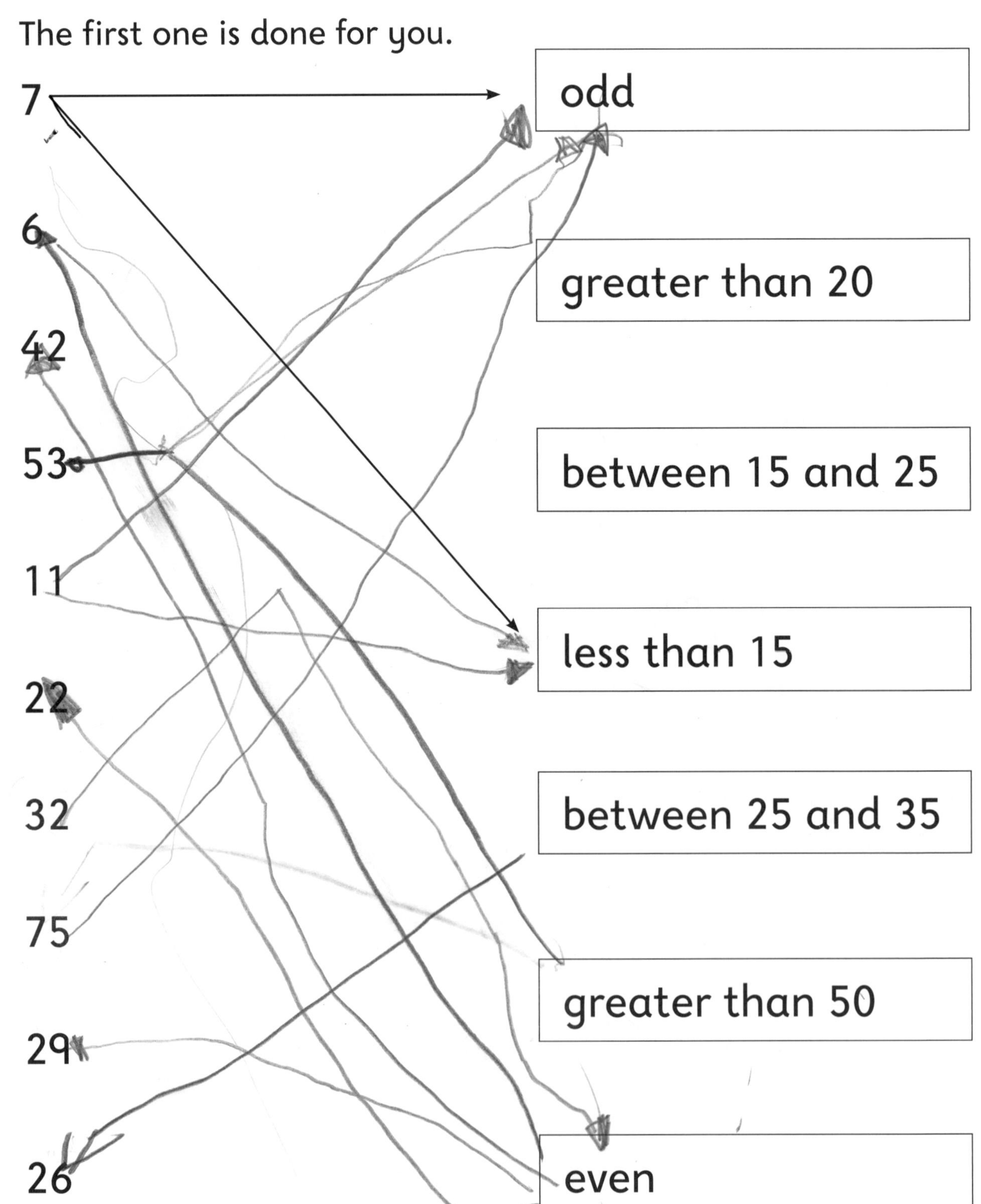

# 3 Number Pairs

## Introduction

There is a mathematical rule called 'commutativity'. This means that it does not matter in which order we add up numbers; we always get the same answer. For example:

7 + 3 = 10 is the same as 3 + 7 = 10

In this unit students will look at pairs of numbers that add together to make 10, 20 and 100. This is useful because these pairs help us with lots of mental calculations.

The students will also begin to round numbers to the nearest multiple of 10.

(The multiples of 10 to 100 are: 10, 20, 30, 40; 50, 60, 70, 80, 90, 100.)

## Ways to help

Number lines are very useful for helping students see which number they should round to. In this unit they start to use empty number lines (lines with no numbers on them). Students can label the ends of the lines with the multiples of 10 that their number is between, to help them see what they should round it to.

Remember that we always round up from 5. So 35 is 40 to the nearest 10.



**Key Words**

number pair; multiple; round; rounding to the nearest 10; round up; round down

# 3A/B Number pairs

## Discover and Explore

Fill in the missing numbers to show some of the number pairs to 20.

| | | | |
|---|---|---|---|
| 7 | + | 13 | = 20 |
| 11 | + | 9 | = 20 |
| 4 | + | 16 | = 20 |
| 14 | + | 6 | = 20 |
| 3 | + | 17 | = 20 |
| 18 | + | 2 | = 20 |
| 1 | + | 19 | = 20 |
| 19 | + | 1 | = 20 |
| 2 | + | 18 | = 20 |
| 17 | + | 3 | = 20 |
| 6 | + | 14 | = 20 |
| 16 | + | 4 | = 20 |
| 9 | + | 11 | = 20 |
| 13 | + | 7 | = 20 |

Tell an adult what you notice about the answers.
Ask them to write what you say.

The answers are all the same.

Write the missing numbers to show the number pairs to 100.

| | | | |
|---|---|---|---|
| 30 | + | 70 | =100 |
| 60 | + | 40 | = 100 |
| 10 | + | 90 | = 100 |
| 80 | + | 20 | = 100 |
| 50 | + | 50 | = 100 |
| 50 | + | 50 | = 100 |
| 20 | + | 80 | = 100 |
| 90 | + | 10 | = 100 |
| 40 | + | 60 | = 100 |
| 70 | + | 30 | = 100 |

Number bonds to make 20.

Number bonds to make 100.

# 3A/B Pairs for rounding

## Discover and Explore

Draw a number line to help you round each number to the nearest 10.
Remember that you always round up from 5. So 65 rounded to the
nearest 10 is 70. The first one is done for you.

**1.** 67 = 70

67 is | 70 | rounded to the nearest 10.

**2.** 53

53 is | 50 | rounded to the nearest 10.

**3.** 78

78 is | 80 | rounded to the nearest 10.

**4.** 16

16 is | 20 | rounded to the nearest 10.

**5.** 88

88 is | 90 | rounded to the nearest 10.

**6.** 7

7 is | 10 | rounded to the nearest 10.

**7.** 98

98 is | 100 | rounded to the nearest 10.

**8.** 75

75 is | 70 | rounded to the nearest 10.

**1.** Write five different number pairs to 10.

One is done for you.

| 2 + 8 = 10 | 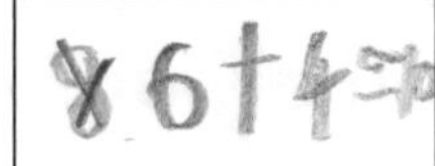 | 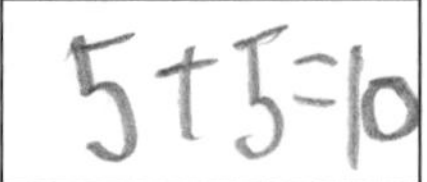 |  |  |

**2.** Write five different number pairs to 20.

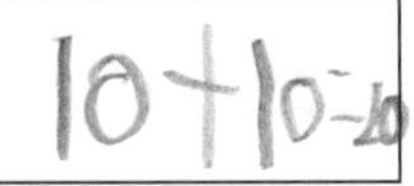  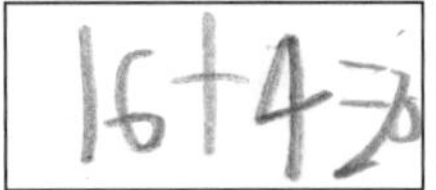 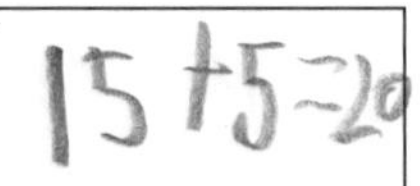 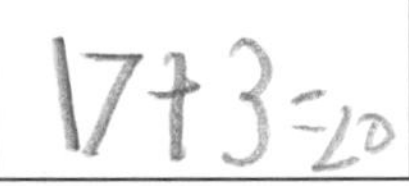

**3.** Write five different number pairs for 100.

 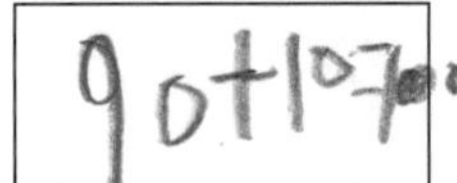 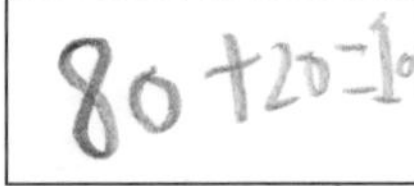  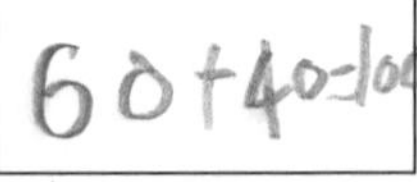

**4.** Round these numbers to the nearest 10.

**a** 28 is _30_ rounded to the nearest 10.

**b** 64 is _60_ rounded to the nearest 10.

**c** 81 is _80_ rounded to the nearest 10.

**d** 15 is _20_ rounded to the nearest 10.

**e** 56 is _60_ rounded to the nearest 10.

**f** 33 is _30_ rounded to the nearest 10.

## Introduction

This unit develops students' understanding of addition and subtraction. They will learn:

- how to use a number line to carry out calculations
- that memorising certain number facts such as number pairs for 10, 20 and 100 can help in mental calculations.

## Ways to help

One way to help is to ask students to explain how they are calculating something. Explaining their strategies for calculating to another person will help them to remember how to carry out a similar calculation in the future.

It is important to avoid telling students incorrect techniques that the students' teacher may have told them are wrong. For example, for the calculation $15 - 7$ we shouldn't say '5 – 7 doesn't go'. (It does go, and the answer is $-2$.) 'Always take the smallest number from the biggest' is also incorrect.

---

**Key Words**

add; addition; subtract; subtraction; take away; difference; count on; count back; number line; number pair; calculation; number sentence

## 4A One more, one less; ten more, ten less

### Discover

You will need two dice for this activity.

Roll both the dice. In the 'Number' column in the table, write the two 2-digit numbers you can make using the numbers on the dice.

Complete each row in the table by writing 10 less, 1 less, 1 more and 10 more than your number.

Use the 100 square to help you.

| 1 | 2 | 3 | 4 | 5 | 6 | 7 | 8 | 9 | 10 |
|---|---|---|---|---|---|---|---|---|---|
| 11 | 12 | 13 | 14 | 15 | 16 | 17 | 18 | 19 | 20 |
| 21 | 22 | 23 | 24 | 25 | 26 | 27 | 28 | 29 | 30 |
| 31 | 32 | 33 | 34 | 35 | 36 | 37 | 38 | 39 | 40 |
| 41 | 42 | 43 | 44 | 45 | 46 | 47 | 48 | 49 | 50 |
| 51 | 52 | 53 | 54 | 55 | 56 | 57 | 58 | 59 | 60 |
| 61 | 62 | 63 | 64 | 65 | 66 | 67 | 68 | 69 | 70 |
| 71 | 72 | 73 | 74 | 75 | 76 | 77 | 78 | 79 | 80 |
| 81 | 82 | 83 | 84 | 85 | 86 | 87 | 88 | 89 | 90 |
| 91 | 92 | 93 | 94 | 95 | 96 | 97 | 98 | 99 | 100 |

| 10 less | 1 less | Number | 1 more | 10 more |
|---|---|---|---|---|
| 24 | 33 | 34 | 35 | 44 |
| 33 | 42 | 43 | 44 | 53 |
| 43 | 52 | 53 | 54 | 63 |
| 53 | 62 | 63 | 64 | 73 |
| 63 | 72 | 73 | 74 | 83 |
| 73 | 82 | 83 | 84 | 93 |
| 83 | 92 | 93 | 94 | 103 |
| 2 | 11 | 12 | 13 | 23 |
| 13 | 22 | 23 | 24 | 33 |
| 34 | 43 | 44 | 45 | 54 |
| 0 | 9 | 10 | 11 | 20 |
| 1 | 10 | 11 | 12 | 22 |

# 4A One more, one less; ten more, ten less

## Explore

You can find all these number patterns in the 100 square.
Can you complete the patterns for 10 less, 1 less, 1 more and 10 more?
The first one is done for you.

| 10 less | | 1 less | | Number pattern | | 1 more | | 10 more | |
|---|---|---|---|---|---|---|---|---|---|
| 44 | 45 | 53 | 54 | 54 | 55 | 55 | 56 | 64 | 65 |
| 54 | 55 | 63 | 64 | 64 | 65 | 65 | 66 | 74 | 75 |

| 10 less | | 1 less | | Number pattern | | 1 more | | 10 more | |
|---|---|---|---|---|---|---|---|---|---|
| 62 | 63 | 71 | 72 | 72 | 73 | 73 | 74 | 82 | 83 |
| 72 | 73 | 81 | 82 | 82 | 83 | 83 | 84 | 92 | 93 |

| 10 less | | 1 less | | Number pattern | | 1 more | | 10 more | |
|---|---|---|---|---|---|---|---|---|---|
| 8 | 9 | 17 | 18 | 18 | 19 | 19 | 20 | 28 | 29 |
| 18 | 19 | 27 | 28 | 28 | 29 | 29 | 30 | 38 | 39 |

| 10 less | | 1 less | | Number pattern | | 1 more | | 10 more | |
|---|---|---|---|---|---|---|---|---|---|
| | | | | 66 | 67 | | | | |
| | | | | 76 | 77 | | | | |

| 10 less | | 1 less | | Number pattern | | 1 more | | 10 more | |
|---|---|---|---|---|---|---|---|---|---|
| | | | | 21 | 22 | | | | |
| | | | | 31 | 32 | | | | |

| 10 less | | 1 less | | Number pattern | | 1 more | | 10 more | |
|---|---|---|---|---|---|---|---|---|---|
| | | | | 78 | 79 | | | | |
| | | | | 88 | 89 | | | | |

### Discover

You will need five dice for this activity or you can roll one dice five times.
Write the numbers in the boxes.

6 + 5 + 4 + 3 + 2 = 20

Add the numbers together and write the answer.
Remember that you can add the numbers in any order.

Circle any pairs of numbers that add to 10. One example is done for you.

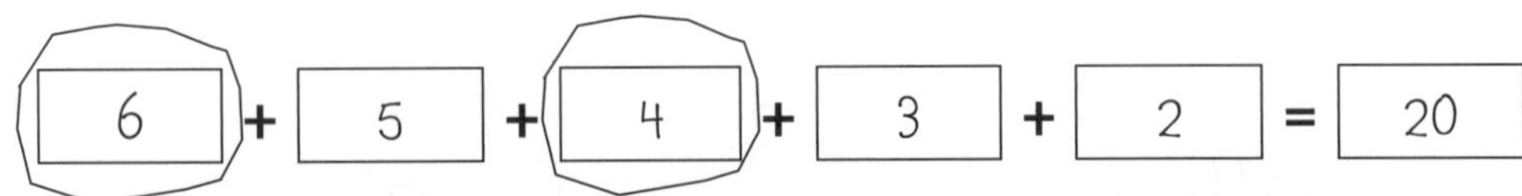

1.  ☐ + ☐ + 3 + 3 + 1 = 15

2.  1 + 1 + 5 + ☐ + 6 = 25

3.  5 + 2 + 2 + 2 + 2 = 13

4.  2 + 1 + 3 + 5 + 1 = 12

5.  6 + 6 + 6 + 6 + 6 = 30

6.  6 + 6 + 6 + 2 + 2 = 22

7.  4 + 4 + ☐ + 4 + 4 = 20

8.  4 + 4 + 5 + 4 + 4 = 21

## Explore

You count forwards on a number line to add and you count backwards to subtract. Draw the jumps on each number line. The first one is done for you.

**1.** 23 + 6 = [ 29 ]

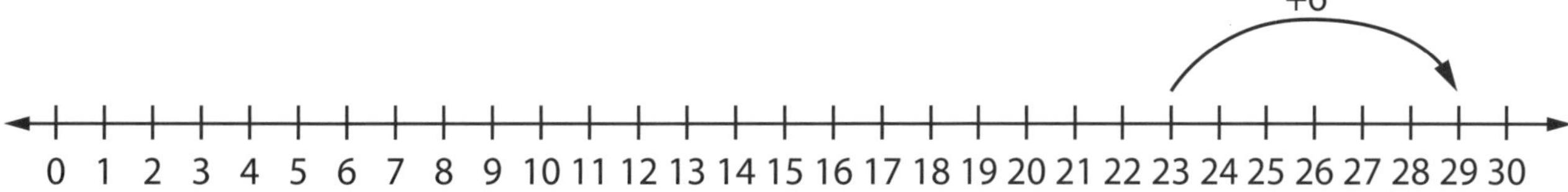


**2.** 16 + 7 = [ ]

**3.** 19 − 4 = [ ]

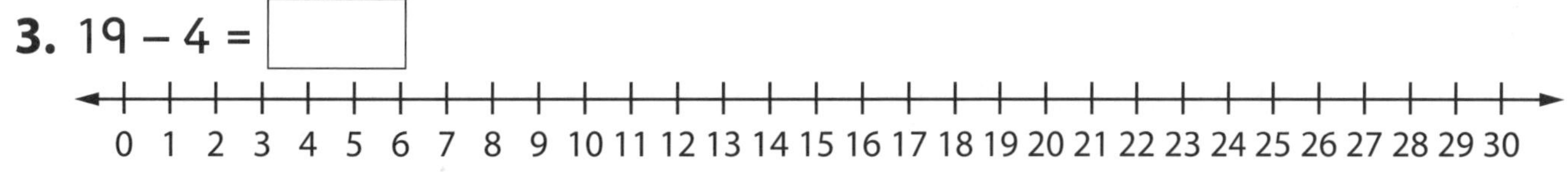


**4.** 27 − 6 = [ ]

**5.** 24 − 4 = [ ]

0 1 2 3 4 5 6 7 8 9 10 11 12 13 14 15 16 17 18 19 20 21 22 23 24 25 26 27 28 29 30

**6.** 18 + 7 = [ ]

0 1 2 3 4 5 6 7 8 9 10 11 12 13 14 15 16 17 18 19 20 21 22 23 24 25 26 27 28 29 30

**7.** 22 + 8 = [ ]

0 1 2 3 4 5 6 7 8 9 10 11 12 13 14 15 16 17 18 19 20 21 22 23 24 25 26 27 28 29 30

**8.** 30 − 8 = [ ]

0 1 2 3 4 5 6 7 8 9 10 11 12 13 14 15 16 17 18 19 20 21 22 23 24 25 26 27 28 29 30

**9.** 12 + 8 = [ ]

0 1 2 3 4 5 6 7 8 9 10 11 12 13 14 15 16 17 18 19 20 21 22 23 24 25 26 27 28 29 30

# 4D Adding two 2-digit numbers

## Discover

You can add 2-digit numbers by partitioning the numbers into tens and ones. This example shows you how.

24 + 32 =

20 + 4 + 30 + 2

You can rearrange this to make it easier to add:

20 + 30 + 4 + 2 = 56

Solve these number sentences.

Partition into tens and ones first. The first one is done for you.

**1.** 28 + 15 = | 20 | + | 10 | + | 8 | + | 5 | = | 43 |

**2.** 16 + 23 = ☐ + ☐ + ☐ + ☐ = ☐

**3.** 24 + 35 = ☐ + ☐ + ☐ + ☐ = ☐

**4.** 35 + 27 = ☐ + ☐ + ☐ + ☐ = ☐

**5.** 18 + 19 = ☐ + ☐ + ☐ + ☐ = ☐

**6.** 34 + 36 = ☐ + ☐ + ☐ + ☐ = ☐

**7.** 28 + 33 = ☐ + ☐ + ☐ + ☐ = ☐

**8.** 33 + 28 = ☐ + ☐ + ☐ + ☐ = ☐

# 4D Adding two 2-digit numbers



Complete the addition grid.
Use the space underneath the grid for your working out.
You can partition the numbers or use a number line.
There may be some doubles or near doubles that will
help you.

| +  | 35 | 24 | 11 | 6 | 18 |
|----|----|----|----|----|----|
| 6  |    |    |    |    |    |
| 27 |    |    |    |    |    |
| 35 |    |    |    |    |    |
| 19 |    |    |    |    |    |
| 20 |    |    |    |    |    |

# 4E Finding the difference

## Discover

You find the difference between two numbers by subtraction.

You can use a number line to help you.
For example, the difference between
29 and 22 is 29 − 22 = 7

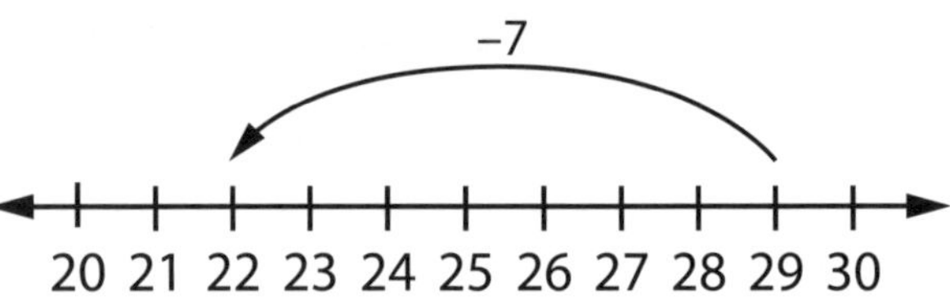

Use the number line to find these differences.
The first one is done for you.

**1.** The difference between 25
and 18 is

| 25 | − | 18 | = | 7 |

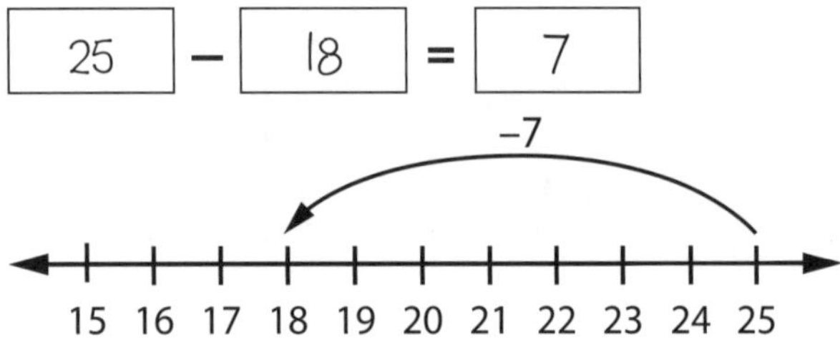

**2.** The difference between 38
and 31 is

| 38 | − | 31 | = | 7 |

**3.** The difference between 42
and 36 is

| 42 | − | 36 | = | 6 |

**4.** The difference between 29
and 21 is

| 29 | − | 21 | = | 8 |

**5.** The difference between 35
and 25 is

| 35 | − | 25 | = | 10 |

**6.** The difference between 14
and 8 is

| 14 | − | 8 | = | 6 |

**7.** The difference between 24
and 18 is

| 24 | − | 18 | = | 6 |

**8.** The difference between 34
and 28 is

| 34 | − | 28 | = | 6 |

# 4E Finding the difference

## Explore

Complete the number sentences with **any** numbers that have the correct difference. The first one is done for you.

**1.** A difference of 6

$$26 - 20 = 6$$

so 26 and 20 have a difference of 6

**2.** A difference of 7

$$25 - 18 = 7$$

so 25 and 18 have a difference of 7.

**3.** A difference of 3

$$24 - 19 = 7$$

so 24 and 19 have a difference of 3.

**4.** A difference of 5

$$10 - 5 = 5$$

so 10 and 5 have a difference of 5.

**5.** A difference of 8

$$16 - 8 = 8$$

so 16 and 8 have a difference of 8.

**6.** A difference of 6

$$34 - 28 = 6$$

so 34 and 28 have a difference of 6.

**7.** A difference of 9

$$36 - 21 = 9$$

so 30 and 21 have a difference of 9.

**8.** A difference of 10

$$20 - 10 = 10$$

so 20 and 10 have a difference of 10.

**9.** A difference of 4

$$8 - 4 = 4$$

so 8 and 4 have a difference of 4.

**10.** A difference of 2

$$4 - 2 = 2$$

so 4 and 2 have a difference of 2.

# 4F Missing numbers

## Discover

Write the missing number in each number sentence.
Tell an adult how you have worked it out. Ask them to write what
you tell them. The first one is done for you.

**1.** 26 + [ 21 ] = 47

I counted on in 10s from 26 to 46. Then I needed 1 more so the answer was 21.

**2.** [ 9 ] + 16 = 25

**3.** 28 + [ 11 ] = 39

**4.** 19 + [ 17 ] = 36

**5.** [ 22 ] + 50 = 72

**6.** [ 12 ] + 25 = 37

# 4F Missing numbers

## Explore

Write the missing number or numbers to make each number sentence correct.

Remember that the total on each side of an equals sign must be the same.

The first one is done for you.

**1.** 11 + [ 37 ] = 21 + 27

**2.** 17 + 13 = 15 + [ 15 ]

**3.** 24 + 18 = [ 14 ] + 28

**4.** 22 + 32 = [ 40 ] + 14

**5.** 7 + 33 = 15 + [ 25 ]

**6.** 24 + [    ] = 14 + [    ]

**7.** 25 + 24 = [ 42 ] + [ 7 ]

**8.** [    ] + 15 = [    ] + 25

Make up a number story for each calculation. Draw a picture to show the story.

You can ask an adult to help you with the writing if you would like to.

The first one is done for you.

**1.** $15 - 7 = 8$

We had 15 cherries. We ate 7.
There were 8 left.

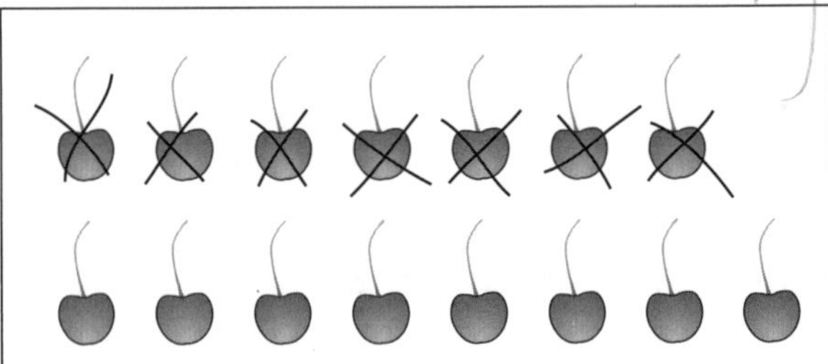

**2.** $12 + 9 = 21$

we had 12 pillos
we bought 9 there
was 41 pillos left

**3.** $22 + 17 = 39$

there was 22
cherries my friend
gave me 17 more there was
39

**4.** $36 - 4 = 32$

there was 36 pillos
4 waere dirty there
was 32 pillos left

**5.** $29 - 23 = 6$

there was 29 plugs
23 was broken there
was 6

# 5   Number Families

## Introduction

In this unit students continue to explore number facts and number patterns, which are useful for carrying out mental calculations. Students have already learned number pairs for 10 and pairs of multiples of 10 which add to 100.

This unit develops and extends their knowledge, so they will learn number pairs for 20 and fact families for multiples of 10 which add to 100. A 'fact family' is a group of facts that we can work out from one single fact. For example, if we know that

1 + 19 = 20

we also know that

19 + 1 = 20

20 − 1 = 19

20 − 19 = 1

## Ways to help

It is helpful to ask students to explain how they are working things out. You can also ask them to work out new facts from facts that they tell you.

If a student tells you that they know that 80 + 20 = 100, ask them what else they can work out from the fact. (20 + 80 = 100, 100 − 20 = 80 and 100 − 80 = 20)

**Key Words**

fact family; number pairs; addition facts; subtraction facts; multiples of 10

# 5A Fact families for number pairs to 20

## Discover and Explore

List all the fact families for number pairs to 20. The first one is done for you.

| Fact families for number pairs to 20 | | | |
|---|---|---|---|
| 1 + 19 = 20 | 19 + 1 = 20 | 20 − 1 = 19 | 20 − 19 = 1 |
| 2 + 18 = 20 | 18 + 2 = 20 | 18 + 2 = 20 | 20 − 18 = 2 |
| 3 + 17 = 20 | 17 + 3 = 20 | 17 + 3 = 14 | 20 − 17 = 3 |
| 4 + 16 = 20 | 16 + 4 = 20 | 16 − 4 = 12 | 20 − 4 = 16 |
| 5 + 15 = 20 | 15 + 5 = 20 | 20 = 15 + 5 | 20 − 5 = 15 |
| 6 + 14 = 20 | 14 + 6 = 20 | 20 = 14 + 6 | 20 − 6 = 14 |
| 7 + 13 = 20 | 13 + 7 = 20 | 20 − 3 = 17 | 20 − 17 = 3 |
| 8 + 12 = 20 | 12 + 8 = 20 | 20 − 12 = 8 | 20 − 8 = 12 |
| 9 + 11 = 20 | 11 + 9 = 20 | 20 − 11 = 9 | 20 − 9 = 11 |
| 10 + 10 = 20 | 10 + 10 = 20 | 20 − 10 = 10 | 20 − 10 = 10 |
| 11 + 9 = 20 | 9 + 11 = 20 | 20 − 11 = 9 | 20 − 9 = 11 |
| 12 + 8 = 20 | 8 + 12 = 20 | 20 − 12 = 8 | 20 − 8 = 12 |
| 13 + 7 = 20 | 7 + 13 = 20 | 20 − 13 = 7 | 20 − 7 = 13 |
| 14 + 6 = 20 | 6 + 14 = 20 | 20 − 14 = 6 | 20 − 6 = 14 |
| 15 + 5 = 20 | 5 + 15 = 20 | 20 − 15 = 5 | 20 − 5 = 15 |
| 16 + 4 = 20 | 4 + 16 = 20 | 20 − 16 = 4 | 20 − 4 = 16 |
| 17 + 3 = 20 | 3 + 17 = 20 | 20 − 17 = 3 | 20 − 3 = 17 |
| 18 + 2 = 20 | 2 + 18 = 20 | 20 − 18 = 2 | 20 − 2 = 18 |
| 19 + 1 = 20 | 1 + 19 = 20 | 20 − 19 = 1 | 20 − 1 = 19 |

# 5B Fact families for multiples of 10 to 100

## Discover and Explore

List all the fact families for multiples of 10 to 100. The first one is done for you.

| Fact families for multiples of 10 to 100 | | | |
|---|---|---|---|
| 10 + 90 = 100 | 90 + 10 = 100 | 100 − 10 = 90 | 100 − 90 = 10 |
| 20 + 80 = 100 | 80 + 20 = 100 | 100 − 20 = 80 | 100 − 20 = 80 |
| 30 + 70 = 100 | 70 + 30 = 100 | 100 − 30 = 70 | 100 − 70 = 30 |
| 40 + 60 = 100 | 60 + 40 = 100 | 100 − 40 = 60 | 100 − 60 = 40 |
| 50 + 50 = 100 | 50 + 50 = 100 | 100 − 50 = 50 | 100 − 50 = 50 |
| 60 + 40 = 100 | 40 + 60 = 100 | 100 − 60 = 40 | 100 − 40 = 60 |
| 70 + 30 = 100 | 30 + 70 = 100 | 100 − 70 = 30 | 100 − 30 = 70 |
| 80 + 20 = 100 | 20 + 80 = 100 | 100 − 80 = 20 | 100 − 20 = 80 |
| 90 + 10 = 100 | 10 + 90 = 100 | 100 − 90 = 10 | 100 − 10 = 90 |

# 5 Review

For each number, write as many facts as you can. Write a fact family of number pairs to 20 or a fact family of multiples of 10 which add to 100. The first two are done for you.

<table>
<tr><td>

**7**

7 + 13 = 20

13 + 7 = 20

20 − 7 = 13

20 − 13 = 7

</td><td>

**40**

40 + 60 = 100

60 + 40 = 100

100 − 40 = 60

100 − 60 = 40

</td></tr>
<tr><td>

**16**

16 + 4 = 20

4 + 16 = 20

20 − 4 = 16

20 − 16 = 14

</td><td>

**70**

70 + 30 = 100

30 + 70 = 100

100 − 30 = 70

100 − 70 = 30

</td></tr>
<tr><td>

**8**

8 + 12 = 20

12 + 8 = 20

20 − 8 = 8

20 − 12 = 12

</td><td>

**100**

90 + 10 = 100

10 + 90 = 100

100 − 90 = 10

100 − 10 = 90

</td></tr>
<tr><td>

**9**

9 + 1 = 10

1 + 9 = 10

10 − 9 = 1

10 − 1 = 9

</td><td>

**90**

90 + 10 = 100

10 + 9 = 100

100 + 10 = 90

100 − 90 = 10

</td></tr>
</table>

# 6 Multiplication and Division

## Introduction

This unit develops the ideas of multiplication and division that were introduced in Stage 1. Students will learn that multiplication is repeated addition. For example:

$3 + 3 + 3 + 3 + 3 = 15$ is the same as $3 \times 5 = 15$

They will also learn about grouping; we can express the same multiplication as '5 groups of 3'. Having 5 groups of 3 things means we have 15 things altogether.

They will develop their understanding of the links between multiplication and division, building on the idea of fact families from the last unit. An example of a fact family for multiplication and division is:

$3 \times 5 = 15$
$5 \times 3 = 15$
$15 \div 5 = 3$
$15 \div 3 = 5$

## Ways to help

Help students to understand dividing by giving them practical sharing activities. For example, ask them to share 15 objects between 5 people. They will then understand that $15 \div 5 = 3$. This also helps them to remember that when we divide we have to split objects into equal groups.

## Discover

Complete this multiplication grid for the 2, 5 and 10 times tables.

If you cannot remember one of the multiplication facts, try to use the facts that you do know to help you work it out. For example:

2 × 8 is double 2 × 4

2 × 4 = 8 so 2 × 8 = 16

| × | 2 | 5 | 10 |
|---|---|---|---|
| 8 | | | |
| 4 | | | |
| 6 | | | |
| 3 | | | |
| 2 | | | |
| 9 | | | |
| 10 | | | |
| 5 | | | |
| 7 | | | |

# 6A/B Twos, fives and tens



You will need a set of 1–9 digit cards for this activity.

| 1 | 2 | 3 | 4 | 5 | 6 | 7 | 8 | 9 |

Pick a digit card at random and write its number in the first column. Then complete the multiplication facts along the row.

Repeat until you have used up all the cards.

| My number | × 2 | × 5 | × 10 |
| --- | --- | --- | --- |
|  |  |  |  |
|  |  |  |  |
|  |  |  |  |
|  |  |  |  |
|  |  |  |  |
|  |  |  |  |
|  |  |  |  |
|  |  |  |  |
|  |  |  |  |

You can ask an adult to help you with the writing if you would like to.

What do you notice about the answers in the '× 5' column?

______________________________________________

What do you notice about the answers in the '× 10' column?

______________________________________________

What do you notice about the answers in the '× 5' column and the '× 10' column?

______________________________________________

# 6C Threes and fours



## Discover

Complete the multiplication grid for the 2, 3 and 4 times tables.

If you cannot remember one of the multiplication facts, try to use the facts that you do know to help you work it out.

| × | 2 | 3 | 4 |
|---|---|---|---|
| 8 | 16 | 24 | 32 |
| 4 | 8 | 12 | 16 |
| 6 | 12 | 18 | 24 |
| 3 | 6 | 9 | 12 |
| 2 | 4 | 6 | 8 |
| 9 | 18 | 27 | 36 |
| 10 | 20 | 30 | 40 |
| 5 | 10 | 15 | 20 |
| 7 | 14 | 21 | 28 |

# 6C Threes and fours

## Explore

You will need a set of 1–9 digit cards for this activity.

| 1 | 2 | 3 | 4 | 5 | 6 | 7 | 8 | 9 |

Pick a digit card at random and write its number in the first column.
Then complete the multiplication facts along the row.

Repeat until you have used up all the cards.

| My number | × 2 | × 3 | × 4 |
| --- | --- | --- | --- |
| 1 | 2 | 3 | 4 |
| 2 | 4 | 6 | 8 |
| 3 | 6 | 9 | 12 |
| 4 | 8 | 12 | 16 |
| 5 | 10 | 15 | 20 |
| 6 | 12 | 18 | 26 |
| 7 | 14 | 21 | 28 |
| 8 | 16 | 24 | 32 |
| 9 | 18 | 27 | 36 |

You can ask an adult to help you with the writing if you would like to.

What do you notice about the answers in the '× 3' column?

goes in 3s

What do you notice about the answers in the '× 4' column?

goes in 4s

What do you notice about the answers in the '× 2' column and the
'× 4' column?

double

# 6D Arrays



Write the fact family for each array. The first one is done for you.

**1.** 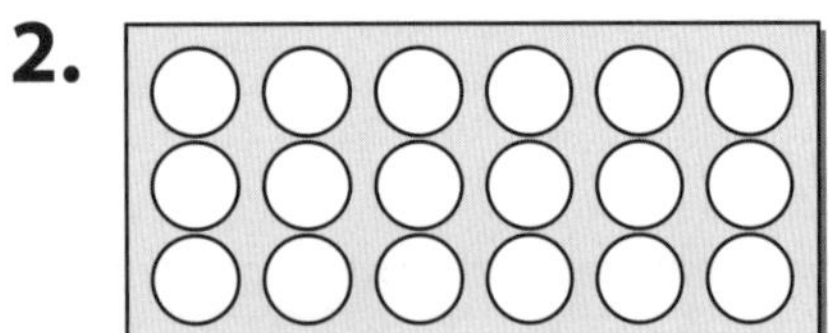

$3 \times 5 = 15$     (3 rows of 5)

$5 \times 3 = 15$     (5 rows of 3)

$15 \div 3 = 5$     (15 objects in 3 rows. 5 objects in each row)

$15 \div 5 = 3$     (15 objects in 5 rows. 3 objects in each row)

**2.**

**3.**

**4.**

**5.**

**6.**

# 6D Arrays



Draw all the different arrays that you can make with 20 counters.

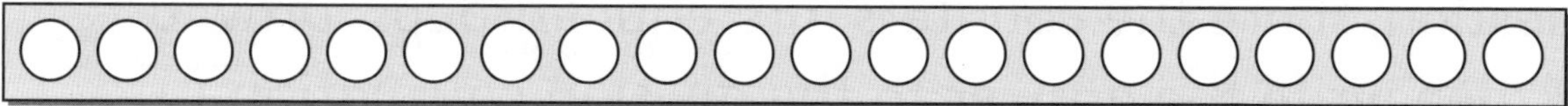

For each array, write the fact family. For example:

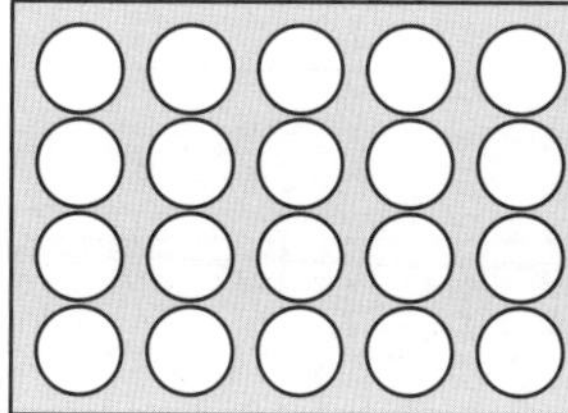

$4 \times 5 = 20$

$5 \times 4 = 20$

$20 \div 4 = 5$

$20 \div 5 = 4$

# 6E Division as grouping



You will need 40 counters, buttons or beads for this activity. For each question, share the counters into equal groups. Draw a picture and then write the answer to the division calculation.

This example shows you how to do it.

$15 \div 3 =$

You need to share 15 counters into groups of 3. How many equal groups of 3 will there be?
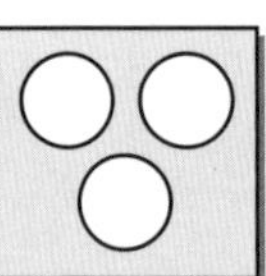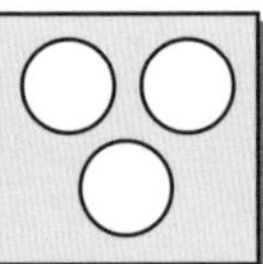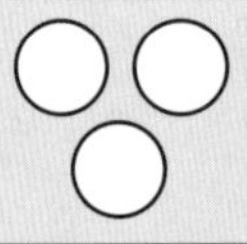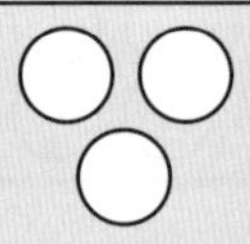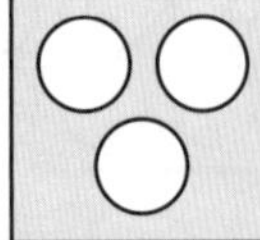

There are 5 equal groups, so $15 \div 3 = 5$

**1.** $32 \div 4 =$ ☐

**4.** $35 \div 5 =$ ☐

**2.** $18 \div 2 =$ ☐

**5.** $36 \div 4 =$ ☐

**3.** $40 \div 10 =$ ☐

**6.** $40 \div 5 =$ ☐

# 6E Division as grouping



Write the multiplication facts for 3. Then write the matching division facts.

You can draw an array to help you. The first one is done for you.

| Fact families for 3 | | | |
|---|---|---|---|
| 1 × 3 = 3 | 3 × 1 = 3 | 3 ÷ 1 = 3 | 3 ÷ 3 = 1 |
| 2 × 3 = | | | |
| 3 × 3 = | | | |
| 4 × 3 = | | | |
| 5 × 3 = | | | |
| 6 × 3 = | | | |
| 7 × 3 = | | | |
| 8 × 3 = | | | |
| 9 × 3 = | | | |
| 10 × 3 = | | | |

Write the multiplication facts for 4. Then write the matching division facts.

You can draw an array to help you. The first one is done for you.

| Fact families for 4 | | | |
|---|---|---|---|
| 1 × 4 = 4 | 4 × 1 = 4 | 4 ÷ 1 = 4 | 4 ÷ 4 = 1 |
| 2 × 4 = | | | |
| 3 × 4 = | | | |
| 4 × 4 = | | | |
| 5 × 4 = | | | |
| 6 × 4 = | | | |
| 7 × 4 = | | | |
| 8 × 4 = | | | |
| 9 × 4 = | | | |
| 10 × 4 = | | | |

# 6F Remainders



This activity introduces the idea of remainders.
You will need 20 counters, buttons or beads.

Use the counters to answer each question. Draw a picture and then write the calculation. The first one is done for you.

**1.** Share 7 apples between 2 people.

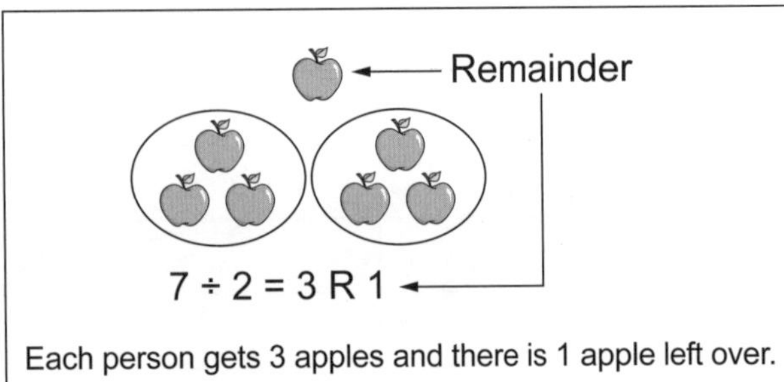


7 ÷ 2 = 3 r 1

**2.** Share 13 counters between 4 people.

**3.** Share 18 counters between 5 people.

**4.** Share 17 counters between 3 people.

**5.** Share 9 counters between 2 people.

**6.** Share 14 counters between 5 people.

# 6F Remainders

## Explore

You will need some counters, buttons or beads for this activity.
Write the calculations and use the counters to help you find the answers.
Some of the calculations will have remainders.
The first one is done for you.

| | ÷ 2 | ÷ 3 | ÷ 4 | ÷ 5 | ÷ 10 |
|---|---|---|---|---|---|
| 15 | 15 ÷ 2 = 7 r 1 | 15 ÷ 3 = 5 | 15 ÷ 4 = 3 r 3 | 15 ÷ 5 = 3 | 15 ÷ 10 = 1 r 5 |
| 18 | | | | | |
| 22 | | | | | |
| 26 | | | | | |
| 16 | | | | | |
| 20 | | | | | |

# 6 Review

Make up a number story for each calculation. Draw a picture to show the story.

You can ask an adult to help you with the writing if you would like to.

The first one is done for you.

**1.** $4 \times 6 = 24$

Fareed has 4 packs of crayons. There are 6 crayons in each pack. So he has 24 crayons altogether.

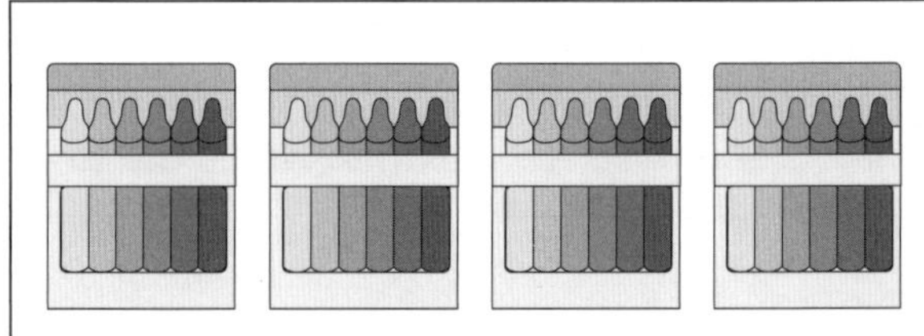

**2.** $3 \times 7 = 21$

**3.** $18 \div 3 = 6$

**4.** $26 \div 5 = 5 \text{ r } 1$

**5.** Make up your own calculation and number story.

# 7 Parts of a Whole

## Introduction

This unit introduces students to fractions. The first fractions that students learn about are halves $\left(\frac{1}{2}\right)$ and quarters $\left(\frac{1}{4}\right)$. The most important thing for students to understand is that when we divide shapes or amounts into fractions, all the parts must be of equal size. This unit makes the link between fractions and division. Students learn that we find one half of an amount by dividing by 2 and one quarter of an amount by dividing by 4.

## Ways to help

Practical experience is always helpful. Practise cutting shapes into halves and quarters and always check that the parts are the same size. Similarly, divide a quantity of objects into halves and quarters by sharing them into equal groups.

**Key Words**

fraction; half $\left(\frac{1}{2}\right)$; quarter $\left(\frac{1}{4}\right)$; three-quarters $\left(\frac{3}{4}\right)$; share; equal parts

## Discover

Find two different ways to split each shape in half.
Colour one half of the shape.
The first one is done for you.

 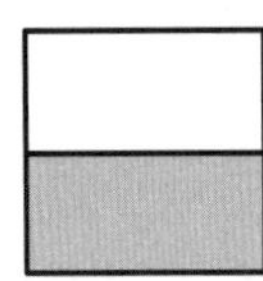 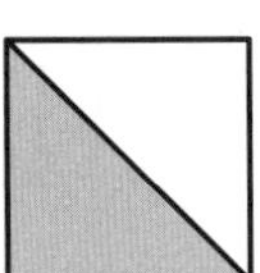

shape          first way          second way

 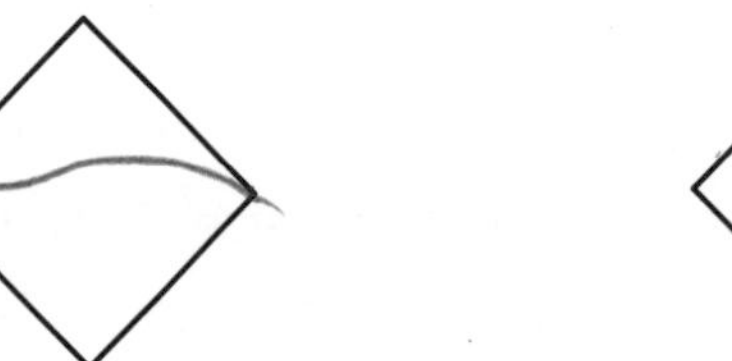

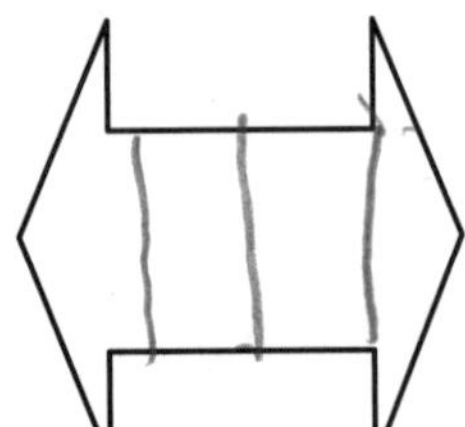 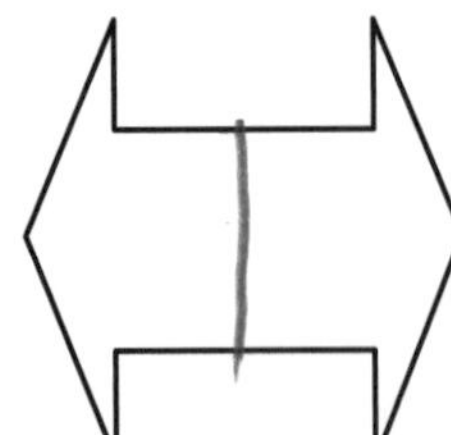 

## Explore

You will need 20 counters, buttons or beads for this activity.

Count out the number of counters shown in the first column. Find half of each number by sharing the counters into two equal groups.

Write how many half the counters is. Draw a picture to show the two halves. The first one is done for you.

| Number of counters | Half ($\frac{1}{2}$) | Picture |
| --- | --- | --- |
| 14 | 7 | |
| 8 | 4 | |
| 12 | 6 | |
| 16 | 8 | |
| 6 | 3 | |
| 20 | 10 | |
| 2 | 1 | |
| 18 | 9 | |
| 10 | 5 | |
| 4 | 2 | |

## Discover

There are different ways to shade a quarter $\left(\frac{1}{4}\right)$ of a shape. Each of these pictures shows $\left(\frac{1}{4}\right)$ of a circle.

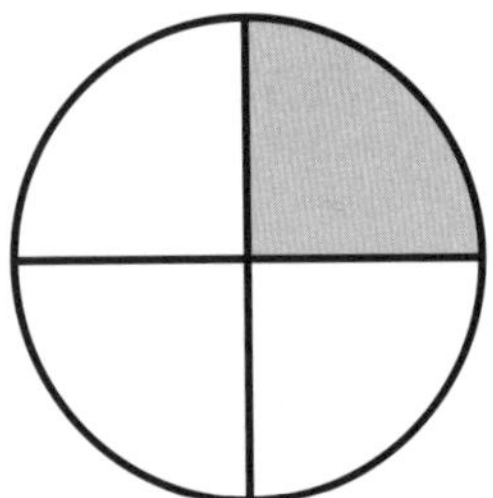 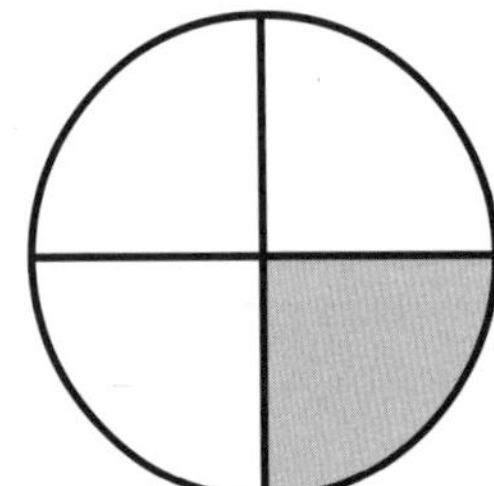 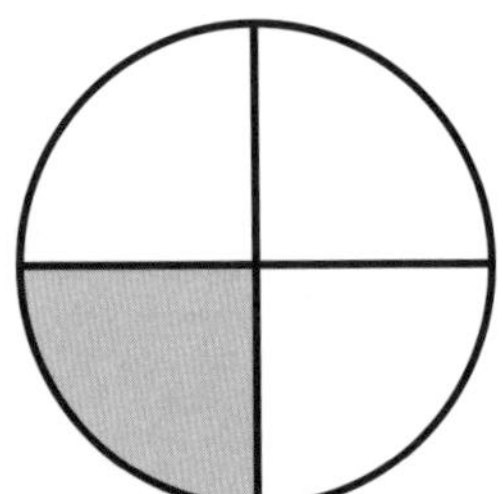 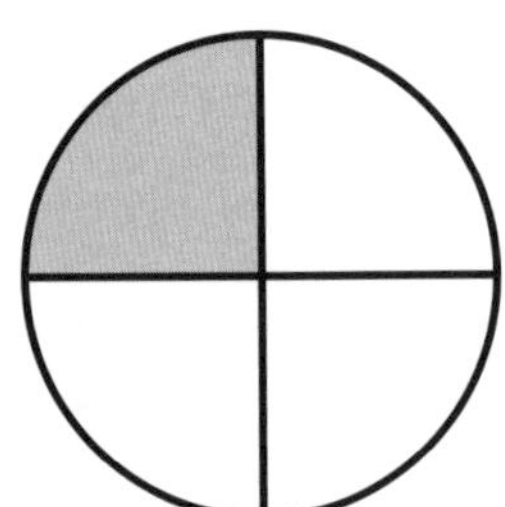

Each of these pictures shows $\frac{1}{4}$ of a square.

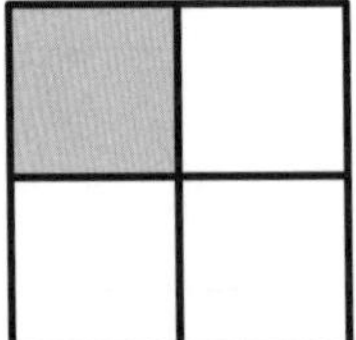 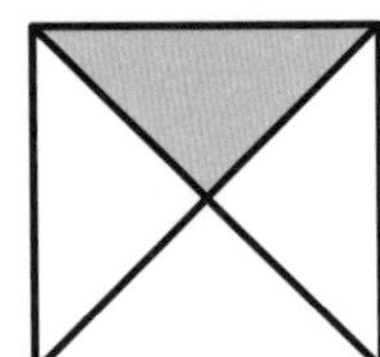 

Find six different ways to shade in $\frac{1}{4}$ of a square.

 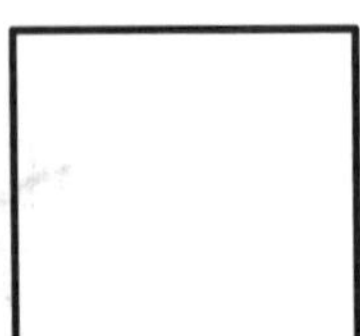 

# 7C Quarter and three-quarters of a shape

## Explore

There are different ways to shade three-quarters $\left(\frac{3}{4}\right)$ of a shape. Each of these pictures shows $\frac{3}{4}$ of a circle.

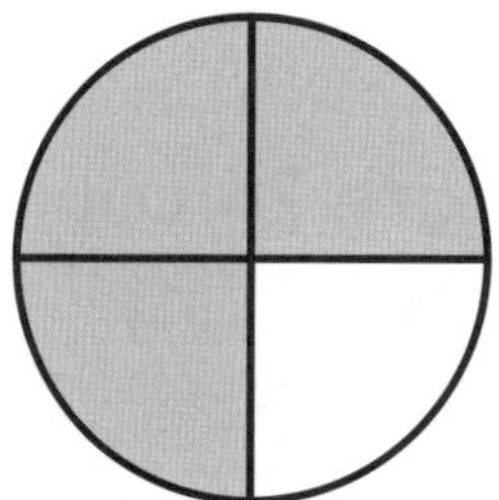 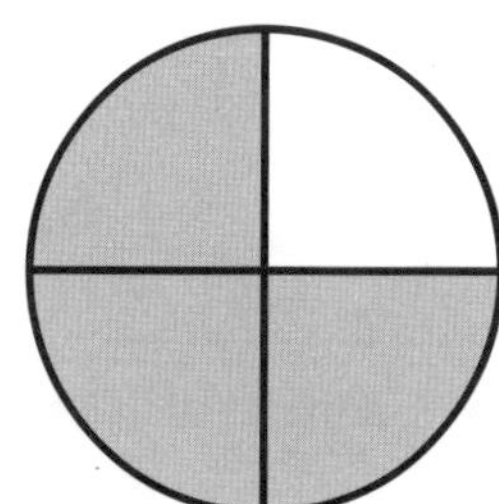

Each of these pictures shows $\frac{3}{4}$ of a square.

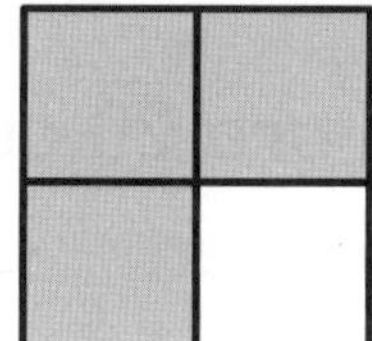 

Find six different ways to shade in $\frac{3}{4}$ of these squares.

  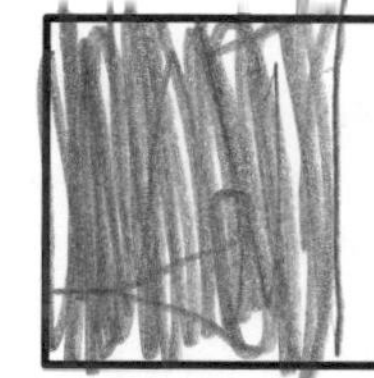

# 7D Quarter of an amount

## Discover

Find $\frac{1}{4}$ of each number of counters. Draw counters inside the quarters of each circle to help you. The first one is done for you.

**1.** There are 12 counters altogether.

$\frac{1}{4}$ of 12 = [ 3 ]

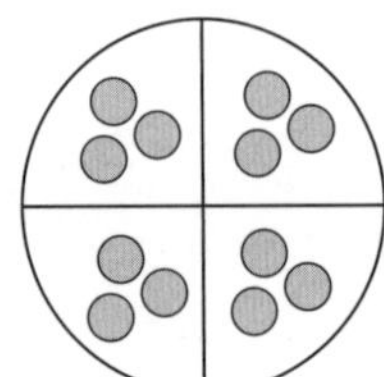

**2.** There are eight counters altogether.

$\frac{1}{4}$ of 8 = [ 2 ]

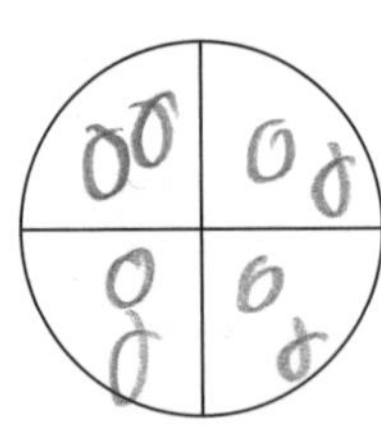

**3.** There are four counters altogether.

$\frac{1}{4}$ of 4 = [ 1 ]

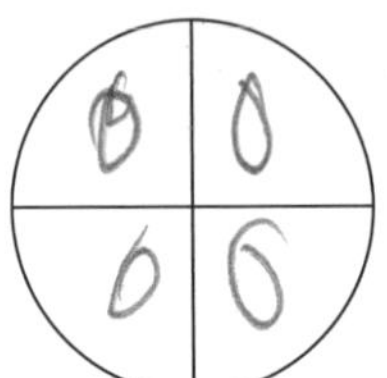

**4.** There are 24 counters altogether.

$\frac{1}{4}$ of 24 = [ 6 ]

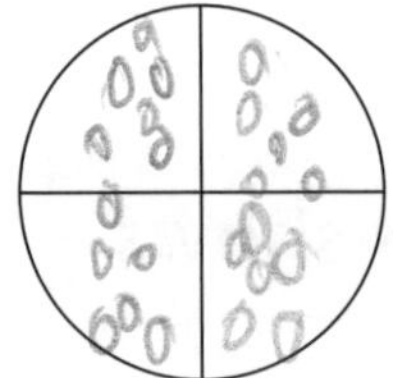

**5.** There are 20 counters altogether.

$\frac{1}{4}$ of 20 = [ 5 ]

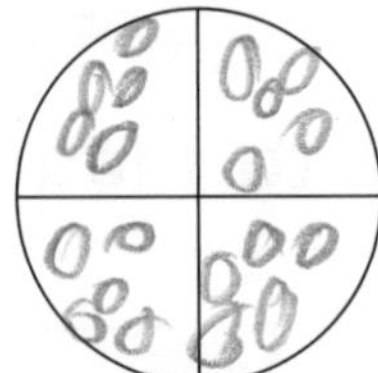

**6.** There are 28 counters altogether.

$\frac{1}{4}$ of 28 = [ 7 ]

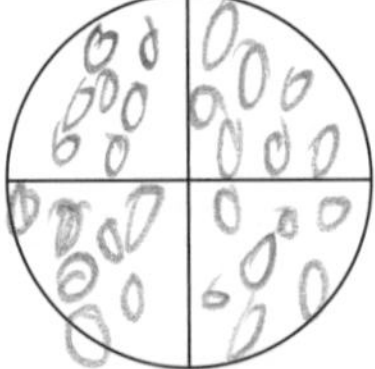

## Explore

Use counters to help you write the missing amounts.

Remember that you find $\frac{1}{4}$ by dividing the counters into four equal groups.

| Number | $\frac{1}{4}$ (Quarter) |
| --- | --- |
| 20 | $\frac{1}{4}$ of 20 = 5 |
| 16 | $\frac{1}{4}$ of 16 = 4 |
| 8 | $\frac{1}{4}$ of 8 = 2 |
| 24 | $\frac{1}{4}$ of 24 = 6 |
| 12 | $\frac{1}{4}$ of 12 = 3 |
| 32 | $\frac{1}{4}$ of 32 = 8 |
| 4 | $\frac{1}{4}$ of 4 = 1 |
| 28 | $\frac{1}{4}$ of 28 = 7 |
| 36 | $\frac{1}{4}$ of 36 = 9 |

Write the missing numbers in these number sentences.

Use counters to help you.

**1.** Half of 20 is $\boxed{10}$

**2.** One-quarter of 20 is $\boxed{5}$

**3.** Three-quarters of 20 is $\boxed{15}$

**4.** $\frac{1}{2}$ of 24 is $\boxed{12}$

**5.** $\frac{1}{4}$ of 24 is $\boxed{6}$

**6.** $\frac{3}{4}$ of 24 is $\boxed{18}$

**7.** $\boxed{\frac{1}{2}}$ of 10 is 5

**8.** $\boxed{\frac{3}{4}}$ of 8 is 6

**9.** $\boxed{\frac{1}{4}}$ of 8 is 2

**10.** Colour in $\frac{1}{4}$, $\frac{1}{2}$ and $\frac{3}{4}$ of this shape.

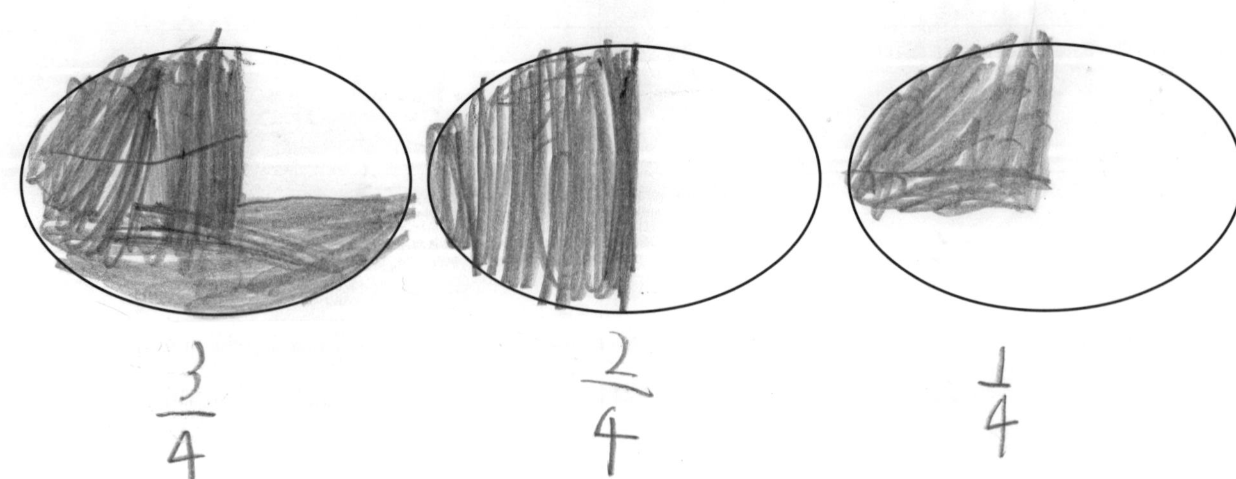

# 8 Shapes Everywhere

## Introduction

This unit introduces students to a wider range of shapes than they met in Stage 1. They will:

- recognise the properties of shapes
- use these properties to describe and classify shapes
- understand that 3-dimensional (3D) shapes have thickness
- understand that 2-dimensional (2D) shapes have no thickness
- continue to develop their understanding of symmetry, which was introduced in Stage 1.

## Ways to help

It is important for students to have practical experience of handling shapes and to see shapes in different orientations.

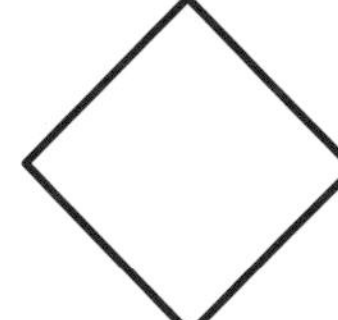

This will allow students to see that all these shapes are squares because they all have four equal sides and four right angles. As in Stage 1, encourage students to notice and talk about shapes in the environment, both at home and out of the home. Take every opportunity you can to name shapes and talk about their properties.

## Key Words

2-dimensional; 2D; square; rectangle; triangle; pentagon; hexagon; side; straight; curved; vertex; vertices

3-dimensional; 3D; face; edge, flat, cube, cuboid, cone; cylinder; sphere; square-based pyramid; symmetrical; line of symmetry; reflective symmetry

# 8A 2D Shapes

## Discover and Explore

At home or outside, try to find objects that are these shapes. You could go on a 'shape-finding walk'.

When you spot an object, draw it in the correct place in the table.
You need to find more than one example of some of the shapes.

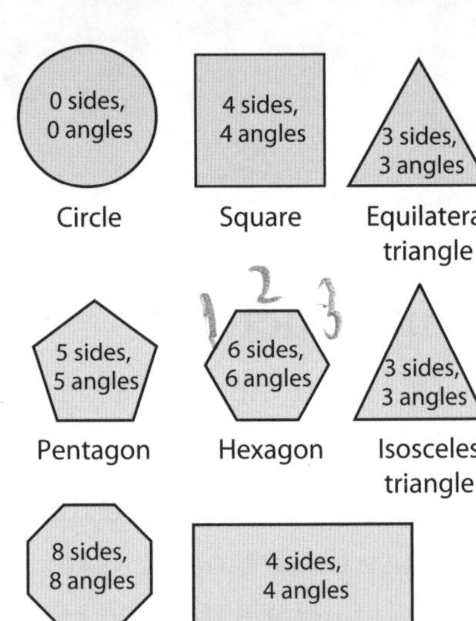



| Shape | Drawing | Shape | Drawing |
|---|---|---|---|
| Circle | | Rectangle | |
| Square | | Pentagon | |
| Triangle | | Pentagon | |
| Triangle | | Hexagon | |
| Triangle | | Hexagon | |

# 8B 3D Shapes

## Discover and Explore

Make a poster of 3D shapes. Find two or three examples of each 3D shape.

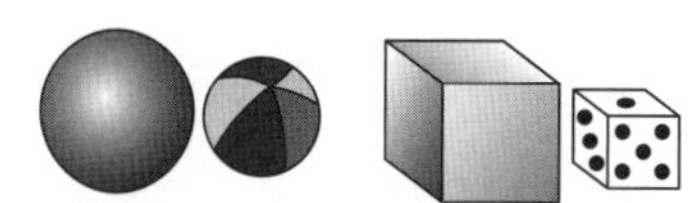

Try to take photographs of objects with these shapes that you see at home or outside. If this is not possible, find and cut out pictures in a magazine or newspaper.

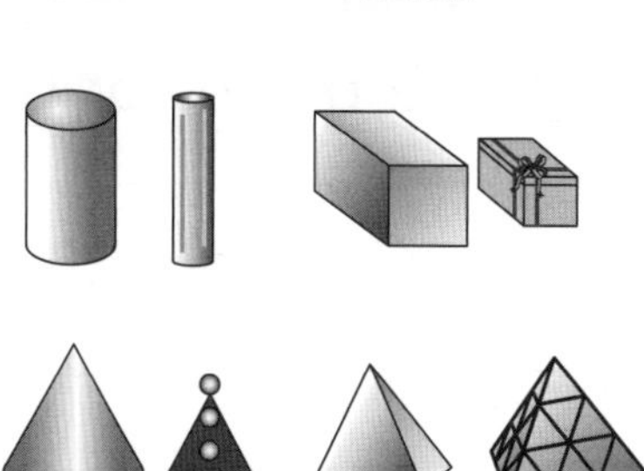

| 3D Shape | Picture |
| --- | --- |
| Sphere | |
| Cube | |
| Cylinder | |
| Cone | |
| Square-based pyramid | |
| Cuboid | |

# 8C/D Sorting shapes

## Discover

Draw shapes to match the properties below. Write the name of each shape.
Two are done for you.

| Properties of shape | Drawing of shape | Name of shape |
| --- | --- | --- |
| 4 equal sides | | square |
| 3 sides<br>Not all equal | | triangle |
| 5 sides<br>Not all equal | | penegon |
| 5 equal sides | | regular pentagon |
| 4 straight sides<br>4 right angles | | ~~cube~~ square |
| 1 curved side<br>No angles | | circle |

# 8C/D Sorting shapes

## Explore

Find pictures of 3D shapes to match the properties below.
Use pictures from magazines or the Internet. Write the
name of each shape.
One is done for you.

| Properties of shape | Picture of shape | Name of shape |
| --- | --- | --- |
| 1 curved face<br>2 flat circular faces | | |
| 4 rectangular faces<br>2 square faces<br>8 vertices | | cuboid |
| 1 curved face<br>No edges or vertices | | sphere |
| 6 square faces<br>8 vertices | | |
| 1 curved face<br>1 flat circular face<br>1 vertex | | |
| 1 square face<br>4 triangular faces<br>5 vertices | | |

## Discover

Draw the lines of symmetry on these shapes. Try to find all the different lines of symmetry.

**1.** A square has 4 lines of symmetry.

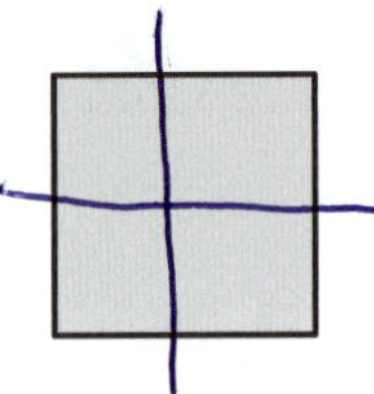

**2.** A rectangle has 2 lines of symmetry.

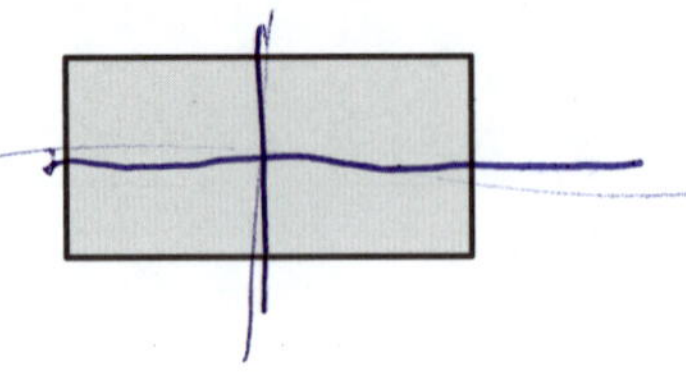

**3.** An equilateral triangle has 3 lines of symmetry.

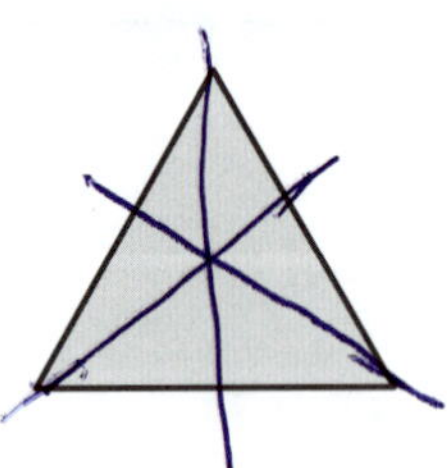

**4.** A regular pentagon has 5 lines of symmetry.

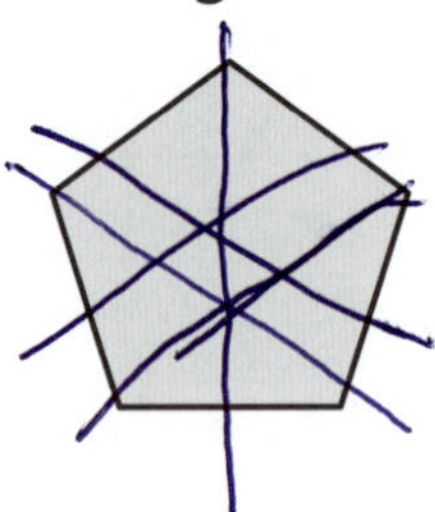

**5.** A regular hexagon has 6 lines of symmetry.

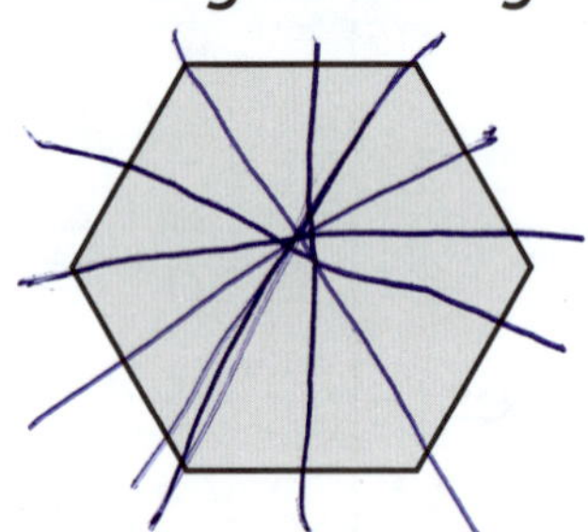

**6.** How many lines of symmetry do you think a circle has? Talk to an adult. Ask them to write what you think.

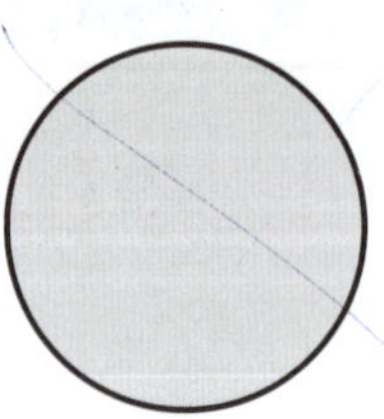

Infinite

# 8E/F Reflective symmetry



## Explore

This pattern has 2 lines of symmetry.

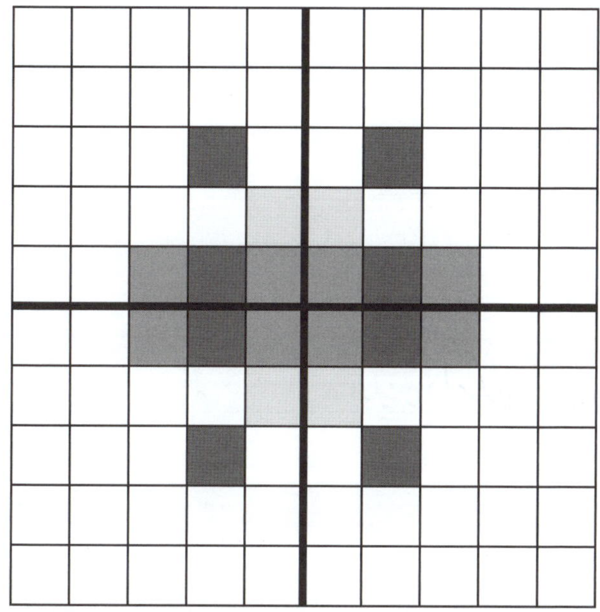

Draw a symmetrical pattern of your own on the grid below. It should have 2 lines of symmetry.

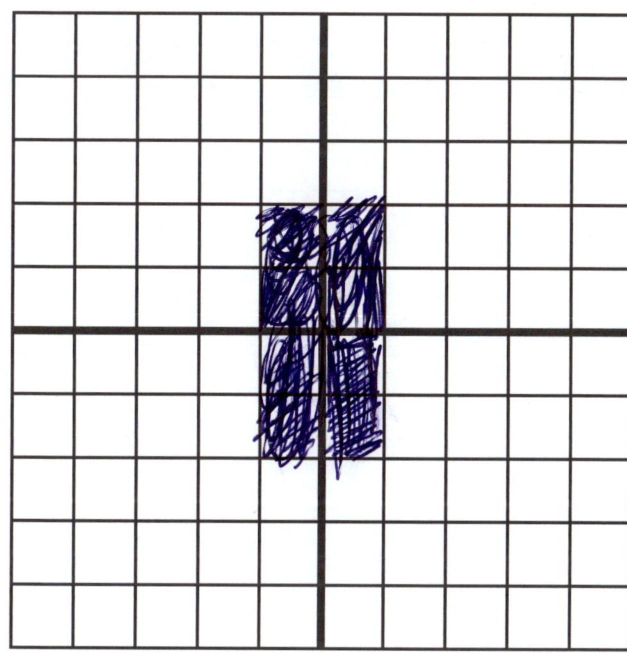

Write as many facts as you can about each shape, including whether it is a 2D or a 3D shape.

You can ask an adult to help you with the writing if you would like to.

**1.** a hexagon

6 sides   6 symmetrys

**2.** a square

4 lines of symetry, 4 sides, equal sides

**3.** a sphere

My name

**4.** a square-based pyramid

3d   bottom is a ▢.

**5.** a cuboid

3d rectangle

**6.** a cube

3d rectange

# 9 Measurements

## Introduction

In Stage 1 students learned that we can use different types of measurement to compare things. In this unit they are introduced to standard units. They learn that we use metres and centimetres to measure lengths; kilograms and grams to measure the mass of objects and litres and millilitres to measure capacity.

## Ways to help

Whenever you measure as a part of day-to-day activities, ask a student to help you. You may need to measure lengths if you are making something. You will need to weigh ingredients and measure amounts of liquids when you are cooking. All of these activities help students to get used to reading scales and give them an understanding of common measures.

It is helpful to look at food and drink containers and packaging with students. When you are shopping, look at the capacities of fruit juice containers and the weights marked on groceries and packets of fruit and vegetables.

## Key Words

metre; centimetre; ruler; tape measure; long; longest; short; shortest; tall; tallest; litre; millilitre; measuring jug; kilogram; gram; heavy; heaviest; light; lightest;

### Discover

You will need a tape measure for this activity.

Find eight objects at home that are shorter than 1 metre in length.
Measure each object. Write its name and its length.

| Object | Length (in cm) |
| --- | --- |
| Mug | 10 cm |
| Book | 20 cm |
| Bottle | 30 cm |
| Phone | 16 cm |
| computer | 38 cm |
| peanut butter | 10 cm |
| Card | 5 cm |
| carrot | 9 cm |

Complete these sentences.

The longest object is __Computer__.

It measures __38__ cm.

The shortest object is __Card__.

It measures __5__ cm.

# 9A/B Measuring in centimetres

## Explore

Draw lines to match these objects to their length.

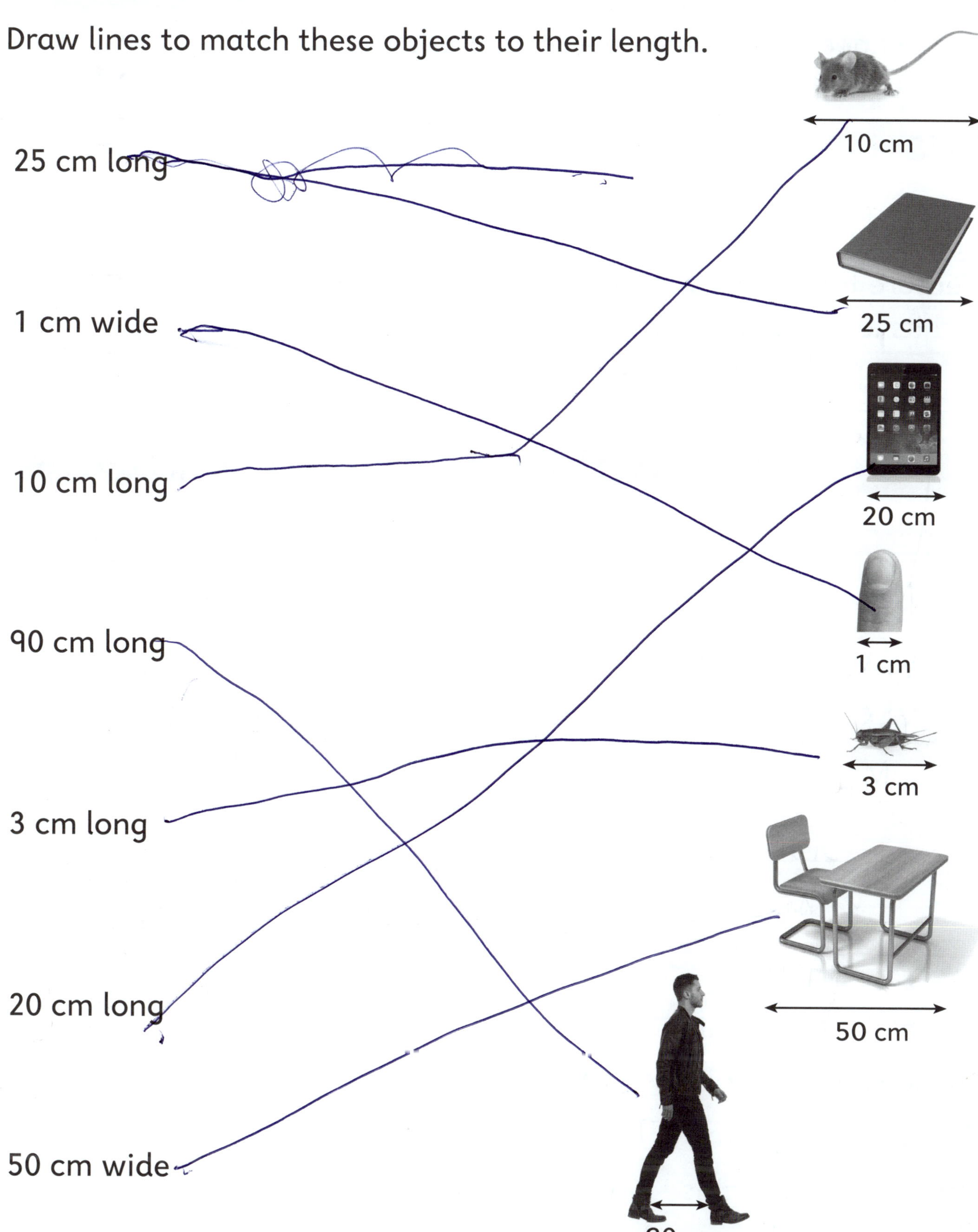

# 9C/D Metres and litres

## Discover

You will need a tape measure for this activity.

Find eight objects at home that are longer than 1 metre.
Measure the length or height of each object. Write the object's name
and its length or height.
An example has been done for you.

| Object | Length or height (in m and cm) |
| --- | --- |
| Door | 2 m 15 cm |
| Me | 146 cm |
| Mother | 1 m 55 cm |
| Dad | 1 m 67 cm |
| Mum | 1 m 53 cm |
| Sister | 1 m 60 cm |
| Fridge | 1.75 m |
| plant | 1.03 m |

Complete these sentences.

The longest object is _____Dad_____.

It measures __1__ m __67__ cm.

The shortest object is _____Plant_____.

It measures __1__ m __3__ cm.

### Explore

You will need a 2-litre measuring jug for this activity.

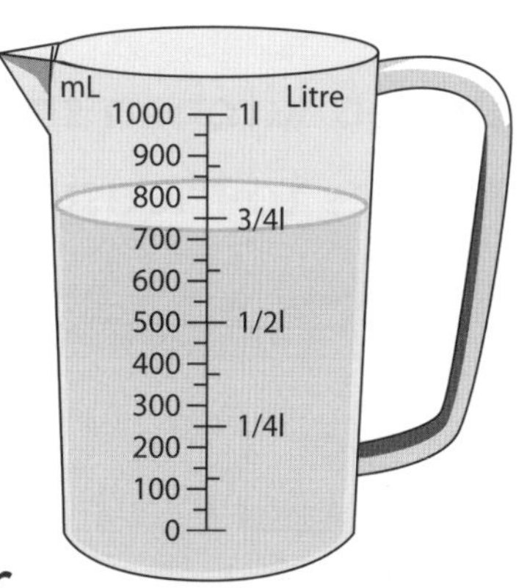


Find six containers at home.
Write the name of each container.
Find out the capacity of each container
by filling it with water and tipping the water
into the measuring jug.
Try to find at least one container that holds more than 1 litre.

| Container | Capacity (in l and ml) |
| --- | --- |
|  |  |
|  |  |
|  |  |
|  |  |
|  |  |
|  |  |
|  |  |

Complete these sentences.

The ________________ has the largest capacity.

It holds ____ l ______ ml.

The ________________________ has the smallest capacity .

It holds ______ ml.

## Discover and Explore

You will need kitchen scales for this activity.
Set the scales to measure in grams.

Find ten objects to weigh. All the objects should weigh less than 1 kg.
Write the name of each object. Then weigh each object and write its weight in grams.
An example is done for you.

| Object | Weight (in g) |
| --- | --- |
| Apple | 157 g |
|  |  |
|  |  |
|  |  |
|  |  |
|  |  |
|  |  |
|  |  |
|  |  |
|  |  |

Complete these sentences.

The heaviest object is _______________________________.

It weighs __________ g.

The lightest object is _______________________________.

It weighs __________ g.

# 9 E/F Money and cooking

## Discover and Explore

Write how much change you would get from $1 for each of the items in the table.
Use the number line to help you. The first one is done for you.

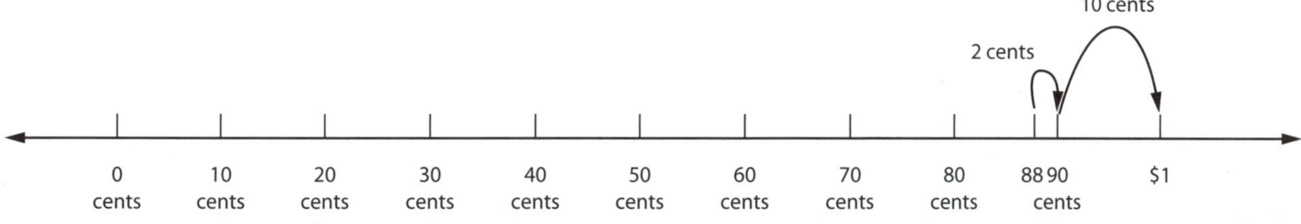


| Item | Cost | Change from $1 |
| --- | --- | --- |
| Book | 88 cents | _12_ cents |
| Fruit | 25 cents | _75_ cents |
| Sweet | 5 cents | _95_ cents |
| Toy | 64 cents | _36_ cents |
| Balloon | 12 cents | _88_ cents |
| Football | 90 cents | _10_ cents |
| Crayons | 75 cents | _25_ cents |
| Juice | 50 cents | _50_ cents |
| Plants | 93 cents | _7_ cents |

Here is a till receipt.

<table>
<tr><td>Product</td><td>Price</td></tr>
<tr><td>1 kg chicken</td><td>$4.00</td></tr>
<tr><td>2 kg potatoes</td><td>$1.00</td></tr>
<tr><td>500 g chick peas</td><td>$0.50</td></tr>
<tr><td>4 limes</td><td>$1.00</td></tr>
<tr><td>10 chillies</td><td>$0.40</td></tr>
</table>

Note:
$0.50 means 50 cents.
$0.40 means 40 cents.

Here is a recipe.

500 g chicken

1 kg potatoes

250 g chick peas

2 limes

5 chillies

What is the *actual* cost of the ingredients used in this recipe?
Use the space on the right-hand side to work it out.

# 10 Geometry

## Introduction

In Stage 1 students learned how to describe the positions of objects. This unit builds on that by teaching students how to describe movements: quarter turns, half turns and movements up, down and forwards on a map grid. They will learn the language of direction, including clockwise and anti-clockwise, to help them to give directions for moving from one place to another. This unit also introduces students to the idea that a right angle is a quarter turn.

## Ways to help

Encourage students to play with jigsaw puzzles, building blocks, toy cars and anything that they can build with or move. Ask them to describe how they are moving and turning these toys and talk about these movements with students. Use language such as 'quarter turn', 'half turn', 'clockwise', 'anti-clockwise', 'up', 'down', 'forwards' and 'backwards'.



**Key Words**

clockwise; anti-clockwise; turn; right angle

### Discover

Make a right-angle measurer by cutting out a circle of card and folding the card twice.

Find six different right angles in your home.
Make a right-angle poster in the space below. Draw the right angles that you find and write the name of each object. You could find pictures in magazines or newspapers too.

computer corner

book

Phone

Board

cards

Radiator

# 10A Turns and right angles

## Explore

Look at the shape in the first column.

Follow the instruction in the second column and draw the new shape in the third column.

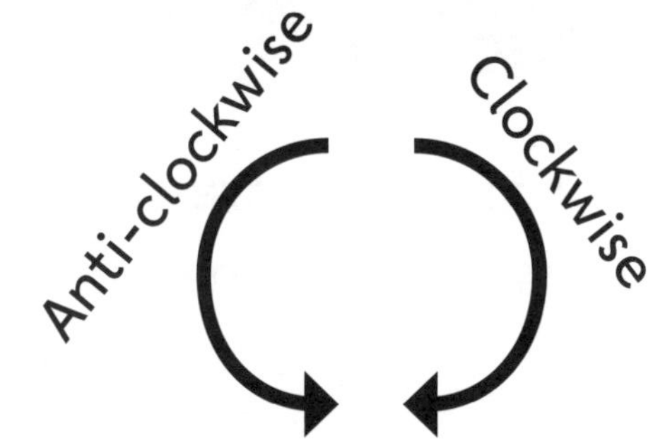


| Shape | Instruction | New shape |
| --- | --- | --- |
| → | quarter turn clockwise | |
| → | quarter turn anti-clockwise | |
| → | half turn | |
| ↔ | half turn | |
| ↔ | quarter turn anti-clockwise | |
| ↔ | quarter turn clockwise | |

## Discover

Imagine you are standing on the grid below. Tell an adult the directions you need to follow to get from one place to another. Ask them to write the directions for you. The first one is done for you. Use some of these words.

| up   down   forwards   quarter turn   clockwise   anti-clockwise |
| --- |

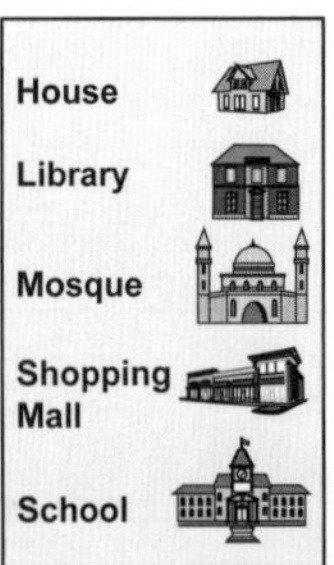



**1.** Go from the mosque to the school.

Up 1 square, quarter turn anti-clockwise, forwards 3 squares.

**2.** Go from the school to the shopping mall.

_______________________________________________

**3.** Go from the house to the shopping mall.

_______________________________________________

**4.** Go from the shopping mall to the school.

_______________________________________________

**5.** Go from the library to the mosque.

_______________________________________________

# 10B Travelling

## Explore

Imagine this grid is a room. Choose a square on the grid to draw:

- a door
- a table
- a bookcase
- a TV
- a sofa

Write five sentences that describe how to move from one object to another in the room.

Each sentence must include at least one turn.

You can ask an adult to help you with the writing if you would like to.

1. _________________________________________________

2. _________________________________________________

3. _________________________________________________

4. _________________________________________________

5. _________________________________________________

Draw a simple map that shows the journey from your house to your school.

Describe the journey from home to school as simply as possible. Write the directions or ask an adult to write them for you.

________________________________________

________________________________________

________________________________________

________________________________________

________________________________________

# 11 Time

## Introduction

Students were introduced to telling the time and days of the week in Stage 1. In this unit they think about measuring time in minutes and seconds and begin to get a sense of how long something takes. They also develop their understanding of telling the time so they can recognise times that are half past the hour.

## Ways to help

Having clocks around the home is invaluable. Try to have both analogue and digital clocks on show. Focus on 'o'clock' and 'half past' times. Try to notice when the time is either an o'clock' time or half past the hour and ask students to tell you what time it is.

analogue clock

digital clock

Write the days of the week on the left-hand side of a large piece of paper. Display this on the wall.

Sunday

Monday

Tuesday

Wednesday

Thursday

Friday

Saturday

Talk about what you do each day.
Write or draw pictures to keep a record.
Keep a monthly calendar on the wall
so that students constantly see the
days of the week and dates of the month.

## Discover and Explore

You will need a stop watch (a timer in seconds) for this activity. Most mobile phones have this function.

Use the stop watch to time yourself doing each activity in the table.
Before you look at the timer, estimate how long you think the activity took.
Write the estimate and then the actual time.

| Activity | Estimated time | Actual time |
| --- | --- | --- |
| Drink a glass of water. | | |
| Walk to the bedroom and back. | | |
| Sit down and stand up 10 times. | | |
| Say the alphabet. | | |
| Count to 100. | | |
| Count backwards from 30 to 0. | | |

## Discover and Explore

Draw the hands on the clocks and write the numbers in the digital clock so that you show the following times.

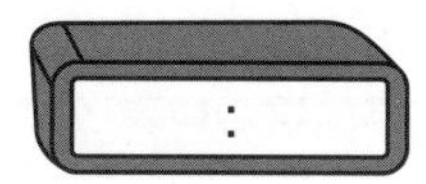

3 o'clock in the morning

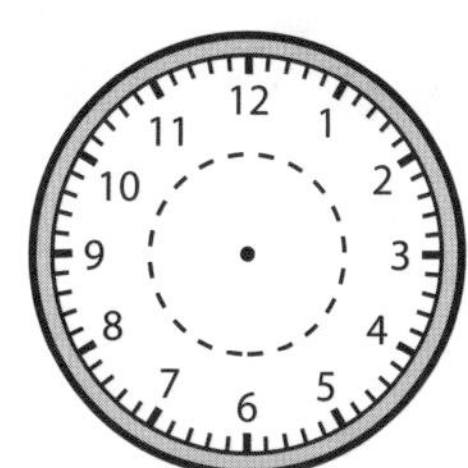

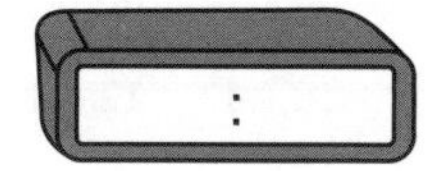

half past 5 in the morning

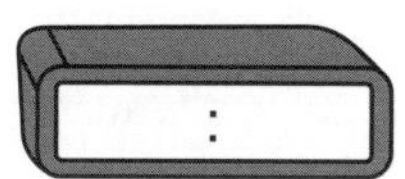

8 o'clock in the morning

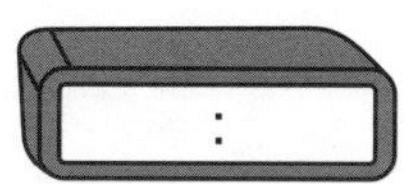

eleven thirty in the morning

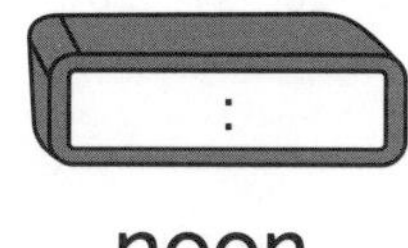

noon

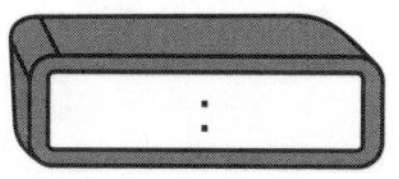

3 o'clock in the afternoon

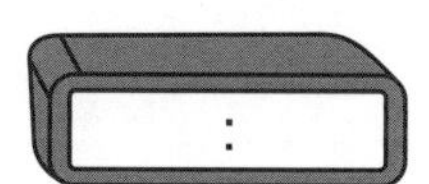

half past 5 in the afternoon

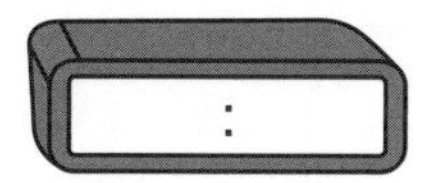

8 o'clock at night

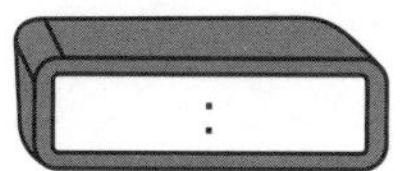

half past eleven at night

### Discover and Explore

# 2015

## January

| Su | Mo | Tu | We | Th | Fr | Sa |
|----|----|----|----|----|----|----|
|    |    |    |    | 1  | 2  | 3  |
| 4  | 5  | 6  | 7  | 8  | 9  | 10 |
| 11 | 12 | 13 | 14 | 15 | 16 | 17 |
| 18 | 19 | 20 | 21 | 22 | 23 | 24 |
| 25 | 26 | 27 | 28 | 29 | 30 | 31 |

## February

| Su | Mo | Tu | We | Th | Fr | Sa |
|----|----|----|----|----|----|----|
| 1  | 2  | 3  | 4  | 5  | 6  | 7  |
| 8  | 9  | 10 | 11 | 12 | 13 | 14 |
| 15 | 16 | 17 | 18 | 19 | 20 | 21 |
| 22 | 23 | 24 | 25 | 26 | 27 | 28 |

## March

| Su | Mo | Tu | We | Th | Fr | Sa |
|----|----|----|----|----|----|----|
| 1  | 2  | 3  | 4  | 5  | 6  | 7  |
| 8  | 9  | 10 | 11 | 12 | 13 | 14 |
| 15 | 16 | 17 | 18 | 19 | 20 | 21 |
| 22 | 23 | 24 | 25 | 26 | 27 | 28 |
| 29 | 30 | 31 |    |    |    |    |

## April

| Su | Mo | Tu | We | Th | Fr | Sa |
|----|----|----|----|----|----|----|
|    |    |    | 1  | 2  | 3  | 4  |
| 5  | 6  | 7  | 8  | 9  | 10 | 11 |
| 12 | 13 | 14 | 15 | 16 | 17 | 18 |
| 19 | 20 | 21 | 22 | 23 | 24 | 25 |
| 26 | 27 | 28 | 29 | 30 |    |    |

## May

| Su | Mo | Tu | We | Th | Fr | Sa |
|----|----|----|----|----|----|----|
|    |    |    |    |    | 1  | 2  |
| 3  | 4  | 5  | 6  | 7  | 8  | 9  |
| 10 | 11 | 12 | 13 | 14 | 15 | 16 |
| 17 | 18 | 19 | 20 | 21 | 22 | 23 |
| 24 | 25 | 26 | 27 | 28 | 29 | 30 |
| 31 |    |    |    |    |    |    |

## June

| Su | Mo | Tu | We | Th | Fr | Sa |
|----|----|----|----|----|----|----|
|    | 1  | 2  | 3  | 4  | 5  | 6  |
| 7  | 8  | 9  | 10 | 11 | 12 | 13 |
| 14 | 15 | 16 | 17 | 18 | 19 | 20 |
| 21 | 22 | 23 | 24 | 25 | 26 | 27 |
| 28 | 29 | 30 |    |    |    |    |

## July

| Su | Mo | Tu | We | Th | Fr | Sa |
|----|----|----|----|----|----|----|
|    |    |    | 1  | 2  | 3  | 4  |
| 5  | 6  | 7  | 8  | 9  | 10 | 11 |
| 12 | 13 | 14 | 15 | 16 | 17 | 18 |
| 19 | 20 | 21 | 22 | 23 | 24 | 25 |
| 26 | 27 | 28 | 29 | 30 | 31 |    |

## August

| Su | Mo | Tu | We | Th | Fr | Sa |
|----|----|----|----|----|----|----|
|    |    |    |    |    |    | 1  |
| 2  | 3  | 4  | 5  | 6  | 7  | 8  |
| 9  | 10 | 11 | 12 | 13 | 14 | 15 |
| 16 | 17 | 18 | 19 | 20 | 21 | 22 |
| 23 | 24 | 25 | 26 | 27 | 28 | 29 |
| 30 | 31 |    |    |    |    |    |

## September

| Su | Mo | Tu | We | Th | Fr | Sa |
|----|----|----|----|----|----|----|
|    |    | 1  | 2  | 3  | 4  | 5  |
| 6  | 7  | 8  | 9  | 10 | 11 | 12 |
| 13 | 14 | 15 | 16 | 17 | 18 | 19 |
| 20 | 21 | 22 | 23 | 24 | 25 | 26 |
| 27 | 28 | 29 | 30 |    |    |    |

## October

| Su | Mo | Tu | We | Th | Fr | Sa |
|----|----|----|----|----|----|----|
|    |    |    |    | 1  | 2  | 3  |
| 4  | 5  | 6  | 7  | 8  | 9  | 10 |
| 11 | 12 | 13 | 14 | 15 | 16 | 17 |
| 18 | 19 | 20 | 21 | 22 | 23 | 24 |
| 25 | 26 | 27 | 28 | 29 | 30 | 31 |

## November

| Su | Mo | Tu | We | Th | Fr | Sa |
|----|----|----|----|----|----|----|
| 1  | 2  | 3  | 4  | 5  | 6  | 7  |
| 8  | 9  | 10 | 11 | 12 | 13 | 14 |
| 15 | 16 | 17 | 18 | 19 | 20 | 21 |
| 22 | 23 | 24 | 25 | 26 | 27 | 28 |
| 29 | 30 |    |    |    |    |    |

## December

| Su | Mo | Tu | We | Th | Fr | Sa |
|----|----|----|----|----|----|----|
|    |    | 1  | 2  | 3  | 4  | 5  |
| 6  | 7  | 8  | 9  | 10 | 11 | 12 |
| 13 | 14 | 15 | 16 | 17 | 18 | 19 |
| 20 | 21 | 22 | 23 | 24 | 25 | 26 |
| 27 | 28 | 29 | 30 | 31 |    |    |

# 11C/D Days and months

## Discover and Explore

Minal has listed her friends' birthdays in the table.

Look at the calendar for 2015 on the opposite page.

Write the day of the week of each of her friends' birthdays in 2015.

Write five dates that are special for you in the table. Write the day of the week of each date in 2015.

| Event | Date | Day of the week |
|---|---|---|
| Saad's birthday | 8 February | |
| Lamah's birthday | 27 March | |
| Aisha's birthday | 22 April | |
| Farhan's birthday | 13 June | |
| Aasif's birthday | 20 July | |
| Nimra's birthday | 4 August | |
| Rimsha's birthday | 8 October | |
| Rayyan's birthday | 30 November | |
| | | |
| | | |
| | | |
| | | |
| | | |

Write the missing days of the week.

Sunday _______________  _______________ Wednesday

_______________ Friday _______________

Write the missing months of the year. Then use a calendar to write how many days there are in each month.

| Month | Number of days |
| --- | --- |
| January |  |
|  |  |
|  |  |
| April |  |
| May |  |
|  |  |
|  |  |
|  |  |
| September | 30 days |
|  |  |
| November |  |
|  |  |

# 12 Handling Data

## Introduction

This unit shows students that we can display data (information) in many different ways. Students will:

- read charts, diagrams and graphs and answer questions about the data shown
- use tally charts, pictograms and block graphs to represent data
- use Carroll diagrams and Venn diagrams to classify (sort) data.

## Ways to help

Collect graphs and charts from newspapers and magazines. Look at these together and talk about what you think they show.

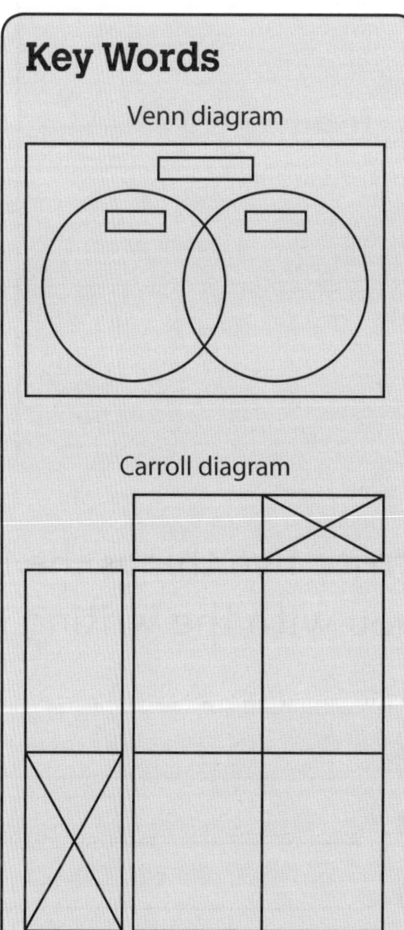

# 12A Block graphs and pictograms

## Discover and Explore

| Day | Number of apples sold |
|---|---|
| Sunday | |
| Monday | |
| Tuesday | |
| Wednesday | |
| Thursday | |
| Friday | |
| Saturday | |

**Key:** = 2 apples

Use the information in the pictogram to draw a bar graph.

Write five things the bar graph tells you. You can ask an adult to help you with the writing if you would like to.

1. _______________________________________________

2. _______________________________________________

3. _______________________________________________

4. _______________________________________________

5. _______________________________________________

# 12B Sorting using Carroll diagrams



**1.** Sort these numbers into the Carroll diagram.

7    12    2    15    26    30    29    14    11    27

|  | Even | Not even |
|---|---|---|
| In the three times table |  |  |
| Not in the three times table |  |  |

**2.** Draw six different 2D shapes in this Carroll diagram. Make sure that you draw at least one shape in each section.

|  | All sides equal | Not all sides equal |
|---|---|---|
| Triangle |  |  |
| Not a triangle |  |  |

# 12C Sorting using Venn diagrams



**1.** Sort these numbers into the Venn diagram. Three numbers are placed for you.

3   5   6   ~~9~~   10   12   ~~15~~   18   20   21   24   ~~25~~   27   30

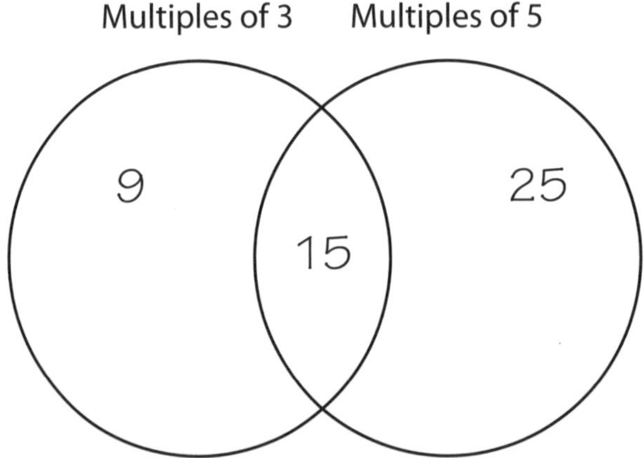

**2.** Draw five shapes in this Venn diagram. Make sure you draw at least one shape in each section.

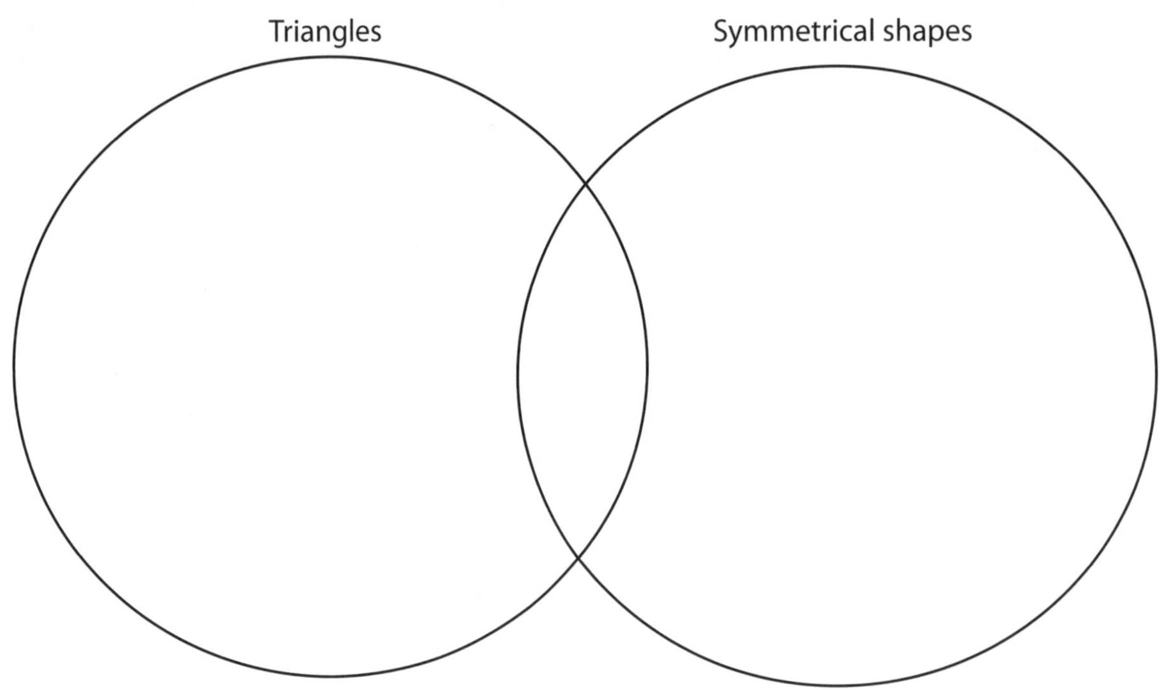

You will need two sets of objects to sort. For example, you could choose food items in a kitchen cupboard, books, kitchen equipment, clothes or toy vehicles.

Write the names of the objects to be sorted. Then write two different ways that you could sort them. For example, one way to sort the items in the food cupboard is into Tins, Boxes, Bottles and Packets.

**1.** I sorted _______________________________________

One way to sort these objects is _______________________

Another way to sort them is _______________________

**2.** I sorted _______________________________________

One way to sort these objects is _______________________

Another way to sort them is _______________________

**3.** Choose one of your sets of objects.

**a** Draw a bar chart showing how many objects there are in each category.

**b** Draw a Venn diagram or a Carroll diagram to show how you sorted your objects.